**Cyber Infrastructure for the Smart
Electric Grid**

Cyber Infrastructure for the Smart Electric Grid

Anurag K. Srivastava
West Virginia University, Morgantown, WV, USA

Venkatesh Venkataramanan
National Renewable Energy Laboratory, Golden, CO, USA

Carl Hauser
Washington State University, Pullman, WA, USA

Registered Office(s)

John Wiley & Sons, Inc., 111 River Street, Hoboken, NJ 07030, USA

John Wiley & Sons Ltd, The Atrium, Southern Gate, Chichester, West Sussex, PO19 8SQ, UK

Editorial Office

The Atrium, Southern Gate, Chichester, West Sussex, PO19 8SQ, UK

For details of our global editorial offices, customer services, and more information about Wiley products visit us at www.wiley.com.

Wiley also publishes its books in a variety of electronic formats and by print-on-demand. Some content that appears in standard print versions of this book may not be available in other formats.

Library of Congress Cataloging-in-Publication Data applied for

Hardback ISBN: 9781119460756

Cover Design: Wiley

Cover Image: © metamorworks/Shutterstock

Set in 9.5/12.5pt STIXTwoText by Straive, Chennai, India

Printed and bound by CPI Group (UK) Ltd, Croydon, CR0 4YY

C9781119460756_241122

Contents

About the Authors *xi*
Acknowledgments *xiii*
Acronyms *xv*

1   **Introduction to the Smart Grid** *1*
1.1   Overview of the Electric Power Grid *1*
1.1.1   Power Grid Operation *6*
1.2   What Can Go Wrong in Power Grid Operation *8*
1.3   Learning from Past Events *9*
1.4   Toward a Smarter Electric Grid *12*
1.5   Summary *13*
1.6   Problems *13*
1.7   Questions *14*
         Further Reading *15*

2   **Sense, Communicate, Compute, and Control in a Secure Way** *17*
2.1   Sensing in Smart Grid *18*
2.1.1   Phase Measurement Unit (PMU) *19*
2.1.1.1 Why Do We Need PMUs? *19*
2.1.1.2 Estimation of Phasors *21*
2.1.1.3 Phasor Calculation *22*
2.1.1.4 Time Signal for Synchronization *22*
2.1.1.5 PMU Data Packets *23*
2.1.1.6 PMU Applications *23*
2.1.2   Smart Meters *24*
2.1.2.1 Communication Systems for Smart Meters *25*
2.2   Communication Infrastructure in Smart Grid *26*

2.3   Computational Infrastructure and Control Requirements in Smart
           Grid   *26*
2.3.1   Control Center Applications   *28*
2.4   Cybersecurity in Smart Grid   *30*
2.4.1   Methods to Provide Cybersecurity for Smart Grids   *31*
2.5   Summary   *31*
2.6   Problems   *31*
2.7   Questions   *33*
           Further Reading   *33*

3   Smart Grid Operational Structure and Standards   *35*
3.1   Organization to Ensure System Reliability   *37*
3.1.1   Regional Entities   *38*
3.2   Smart Grid Standards and Interoperability   *39*
3.3   Operational Structure in the Rest of the World   *40*
3.4   Summary   *41*
3.5   Problems   *41*
3.6   Questions   *42*
           Further Reading   *42*

4   Communication Performance and Factors that Affect It   *45*
4.1   Introduction   *45*
4.2   Propagation Delay   *47*
4.3   Transmission Delay   *47*
4.4   Queuing Delay and Jitter   *49*
4.5   Processing Delay   *51*
4.6   Delay in Multi-hop Networks   *51*
4.7   Data Loss and Corruption   *52*
4.8   Summary   *53*
4.9   Exercises   *53*
4.10   Questions   *56*
           Further Reading   *56*

5   Layered Communication Model   *57*
5.1   Introduction   *57*
5.1.1   OSI and TCP/IP Models   *59*
5.2   Physical Layer   *60*
5.3   Link Layer: Service Models   *61*
5.3.1   Ethernet   *62*
5.3.1.1   Link Virtualization   *63*

5.4   Network Layer: Addressing and Routing   *64*
5.4.1   IP Addressing   *66*
5.4.2   Routing   *68*
5.4.3   Broadcast and Multicast   *68*
5.5   Transport Layer: Datagram and Stream Protocols   *70*
5.5.1   UDP   *72*
5.5.2   TCP   *73*
5.6   Application Layer   *75*
5.7   Glue Protocols: ARP and DNS   *76*
5.7.1   DNS   *77*
5.8   Comparison Between OST and TCP/IP Models   *78*
5.9   Summary   *78*
5.10   Problems   *79*
5.11   Questions   *80*
          Further Reading   *80*

6   Power System Application Layer Protocols   *81*
6.1   Introduction   *81*
6.2   SCADA Protocols   *82*
6.2.1   DNP3 Protocol   *83*
6.2.2   IEC 61850   *86*
6.3   ICCP   *87*
6.4   C37.118   *87*
6.5   Smart Metering and Distributed Energy Resources   *89*
6.5.1   Smart Metering   *89*
6.5.2   Distributed Energy Resources (DERs)   *91*
6.6   Time Synchronization   *92*
6.7   Summary   *92*
6.8   Problems   *93*
6.9   Questions   *94*
          Further Reading   *94*

**7   Utility IT Infrastructures for Control Center and
          Fault-Tolerant Computing**   *95*
7.1   Conventional Control Centers   *95*
7.2   Modern Control Centers   *97*
7.3   Future Control Centers   *98*
7.4   UML, XML, RDF, and CIM   *99*
7.4.1   UML   *100*
7.4.2   XML and RDF   *102*

7.4.3   CIM (IEC 6170)   *103*
7.4.4   IEC 61850   *103*
7.5   Basics of Fault-Tolerant Computing   *105*
7.6   Cloud Computing   *107*
7.7   Summary   *109*
7.8   Problems   *110*
7.9   Questions   *111*
         Further Reading   *111*

8   **Basic Security Concepts, Cryptographic Protocols, and**
             Access Control   *113*
8.1   Introduction   *113*
8.2   Basic Cybersecurity Concepts and Threats to Power Systems   *113*
8.2.1   Threats, Vulnerabilities, and Risks, What Is the Difference?   *113*
8.2.2   Threats   *114*
8.2.3   Vulnerabilities   *114*
8.2.4   Risk   *115*
8.3   CIA Triad and Other Core Security Properties   *116*
8.3.1   Privacy and Consumer Data   *117*
8.3.2   Encryption and Authentication   *117*
8.3.2.1 Kerckhoffs's versus Kirchoff's Law (Fundamental Cryptographic
         Principles and Threats)   *118*
8.3.2.2 Symmetric Key Encryption   *120*
8.3.2.3 Asymmetric Key   *121*
8.4   Introduction to Encryption and Authentication   *123*
8.4.1   Message Authentication Codes (MACs)   *123*
8.4.2   Digital Signatures   *124*
8.4.3   Certificates   *125*
8.5   Cryptography in Power Systems   *127*
8.5.1   IEC 62351   *128*
8.5.2   DNP3 Secure Authentication (SA)   *129*
8.6   Access Control   *131*
8.6.1   RBAC in IEC 62351   *131*
8.7   Summary   *133*
8.8   Problems   *133*
8.9   Questions   *134*
         Further Reading   *134*

9   **Network Attacks and Protection**   *135*
9.1   Attacks to Network Communications   *135*
9.1.1   Denial-of-Service (DoS) Attack   *135*

9.1.1.1   Flooding   *136*
9.1.1.2   Malformed Packet   *137*
9.1.1.3   Reflection   *137*
9.1.1.4   DDoS   *138*
9.1.2   Spoofing   *138*
9.1.2.1   ARP Spoofing   *139*
9.1.2.2   Other Spoofing   *139*
9.2   Mitigation Mechanisms Against Network Attacks   *140*
9.2.1   Network Protection Through Security Protocols   *140*
9.2.1.1   TLS   *141*
9.2.1.2   IPsec   *143*
9.3   Network Protection Through Firewalls   *144*
9.4   Intrusion Detection   *145*
9.4.1   Anomaly-Based Detection   *147*
9.4.2   Signature-Based Detection   *147*
9.5   Summary   *148*
9.6   Problems   *149*
9.7   Questions   *150*
Further Reading   *150*

10   **Vulnerabilities and Risk Management**   *151*
10.1   System Vulnerabilities   *151*
10.1.1   Software Vulnerabilities   *152*
10.1.2   Hardware and Side-Channel Vulnerabilities   *155*
10.1.3   Social Engineering   *155*
10.1.4   Malware   *156*
10.1.5   Supply Chain   *158*
10.2   Security Mechanisms: Access Control and Malware Detection   *159*
10.2.1   Access Control   *159*
10.2.2   Malware Detection   *160*
10.3   Assurance and Evaluation   *161*
10.3.1   Port Scanning   *161*
10.3.2   Network Monitoring   *162*
10.3.3   Network Policy Analysis   *163*
10.3.4   Vulnerability Scanning   *163*
10.3.5   Continuous Monitoring   *163*
10.3.6   Security Assessment Concerns   *165*
10.3.7   Software Testing   *165*
10.3.8   Evaluation   *166*
10.4   Compliance: Industrial Practice to Implement NERC CIP   *167*
10.5   Summary   *167*

10.6   Problems   *167*
10.7   Questions   *168*
         Further Reading   *169*

11   **Smart Grid Case Studies**   *171*
11.1   Smart Grid Demonstration Projects   *171*
11.2   Smart Grid Metrics   *173*
11.3   Smart Grid Challenges: Attack Case Studies   *174*
11.3.1   Stuxnet   *175*
11.3.2   Ukraine Attack   *176*
11.4   Mitigation Using NIST Cybersecurity Framework   *178*
11.5   Summary   *180*
11.6   Problems   *180*
11.7   Questions   *181*
         Further Reading   *181*

         Index   *183*

About the Authors

 **Anurag K. Srivastava, PhD,** is the Raymond J. Lane Professor and Chairperson of the Lane Department of Computer Science and Electrical Engineering in the Benjamin M. Statler College of Engineering and Mineral Resources at West Virginia University. He is the director of the Smart Grid Resiliency and Analytics Lab (SGREAL) and an IEEE Fellow.

 **Venkatesh Venkataramanan, PhD,** is a Researcher at the National Renewable Energy Laboratory, working on cyber-physical systems. He was previously with Washington State University and Massachusetts Institute of Technology.

 **Carl Hauser, PhD,** is emeritus faculty in Computer Science at Washington State University. He received his PhD from Cornell University. Following 20 years in industry at IBM Research and Xerox Research, he joined WSU where he conducted research on communications and cybersecurity for electric grid operations.

Acknowledgments

Authors are thankful to students who were brave enough to take the team-taught course offered at the Washington State University. Students shaped up the course material development process over multiple offerings. Authors are also thankful to the US Department of Energy and the Power System Engineering Research Center (PSERC) for supporting the course development. We acknowledge the support from our colleagues including Dr. Adam Hahn, Prof. David Bakken, Dr. Min Sik Kim, and Prof. Anjan Bose.

Acronyms

ASTA	Arrivals See Time Averages
BHCA	Busy Hour Call Attempts
BR	Bandwidth Reservation
b.u.	bandwidth unit(s)
CAC	Call/Connection Admission Control
CBP	Call Blocking Probability(-ies)
CCS	Centum Call Seconds
CDTM	Connection-Dependent Threshold Model
CS	Complete Sharing
DiffServ	Differentiated Services
EMLM	Erlang Multirate Loss Model
erl	The Erlang unit of traffic-load
FIFO	First in–First out
GB	global balance
GoS	Grade of Service
ICT	Information and Communication Technology
IntServ	Integrated Services
ITU-T	International Telecommunication Unit – Standardization sector
IP	Internet Protocol
LIFO	Last in–First out
LHS	left hand side
LB	local balance
MMPP	Markov Modulated Poisson Process
MPLS	Multiple Protocol Labeling Switching
MRM	multi-retry model
MTM	multi-threshold model
PASTA	Poisson Arrivals See Time Averages
pdf	probability density function
PDF	probability distribution function

PFS	product form solution
QoS	quality of service
RED	random early detection
r.v.	random variable(s)
RLA	reduced load approximation
RHS	right-hand side
SIRO	service in random order
SRM	single-retry model
STM	single-threshold model
TH	Threshold(s)
TCP	Transport Control Protocol
UDP	User Datagram Protocol

1

Introduction to the Smart Grid

The power grid has been evolving from a physical system to a "cyber-physical" system to sense, communicate, compute, and control with enhanced digitalization. The cyber-physical smart grid includes components from the physical power system, digital devices, and the associated communication infrastructure. To realize the vision of the smart grid, massive amounts of data need to be transferred from the field devices to the control devices or to the control centers. As more optimal algorithms are deployed for best possible control at a faster time scale, the communication infrastructure becomes critical to provide the required inputs. At the same time, increased number of "smart" devices in the grid also increase the attack surface for potential cyber attacks. It is necessary to study the power system's exposure to risks and vulnerabilities in the associated cyber system.

1.1 Overview of the Electric Power Grid

The electric power grid can be defined as the entire apparatus of wires and machines that connects the sources of electricity with the customers. A power grid is generally divided into four major components as shown in Figure 1.1:

1. Generation
2. Transmission
3. Distribution
4. Loads

Electricity was first generated, sold, and distributed locally in 1870s via direct current (DC) circuits over very small distances. As the demand for electricity became more widespread, the cost of construction and distribution of local generation and DC circuits to carry the power over long distances became prohibitively expensive. Hence, alternating current (AC) generation, transmission, and distribution became the standard that is used to this day. However, the

Cyber Infrastructure for the Smart Electric Grid, First Edition.
Anurag K. Srivastava, Venkatesh Venkataramanan, and Carl Hauser.
© 2023 John Wiley & Sons Ltd. Published 2023 by John Wiley & Sons Ltd.

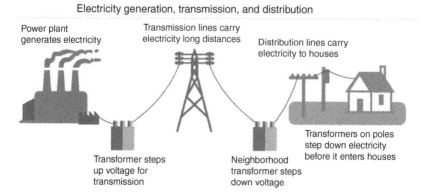

Electricity generation, transmission, and distribution

Figure 1.1 Major components of the power grid. Source: Energy Information Administration (EIA), public domain.

infrastructure of the power grid is getting older – the average age of a transformer is greater than 50 years old and has already exceeded its expected lifetime. The electric grid faces several problems, including a problem with the oncoming retirement of at least 5% of the workforce and one of the lowest R&D expenditure as compared to other critical infrastructures.

The situation is getting better, however, with increasing interest in national security and acknowledgment of the critical role that the power grid plays in the overall quality of life. In a full circle, localized generation using distributed energy resources (DERs) is making a comeback, with a combination of both AC and DC systems. Today's generation systems are a combination of different types of sources – including fossil fuels, natural gas, renewable resources, and nuclear energy. These generation systems are often located in remote areas for ease of doing business and for environmental reasons.

The power that is generated at the generating stations is brought to the consumers by a complex network of transmission lines. The North American power grid comprises of four major interconnections as shown in Figure 1.2:

1. Western interconnection
2. Eastern interconnection
3. Quebec interconnection
4. Electricity Reliability Council of Texas (ERCOT) interconnection

These interconnections are zones in which the electric utilities are electrically tied together, indicating that the areas are synchronized to the same frequency and power can flow freely in that area. The interconnections operate nearly independently of each other except for some high-voltage direct current (HVDC)

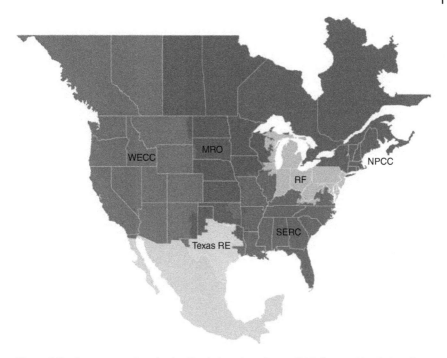

Figure 1.2 Interconnections in the North American Power Grid. Source: North American Energy Reliability Corporation (NERC), public domain.

interconnections between them. DC converter substations enable the synchronized transfer of power across interconnections regardless of the operating frequency as DC power is non-phase dependent.

The flow of electricity is instantaneous, indicating that the power that is being consumed is also being simultaneously generated. Commercially viable mechanisms for storing electricity for longer duration do not exist currently; hence, the power plants and the grid are constantly operating. The structure of the flow of electricity is illustrated in Figure 1.3, which shows the critical nature of the transmission system in bringing electricity from the generating plants to the customer's use.

Power demand constantly fluctuates throughout the day depending on consumer behavior. There are various factors that create this changing behavior, including population density, work schedules, weather, and other activities. In addition, special activities that involve a large number of people also have to be considered, such as big sporting events or an impending weather event over a large area. Figure 1.4 shows a typical daily "load" curve as it is referred to, which shows how the electric load varies across a day depending on the activities

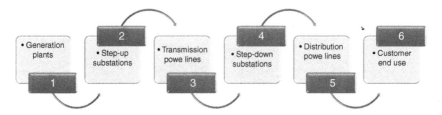

Figure 1.3 Structure of electricity flow from generating stations to the consumer.

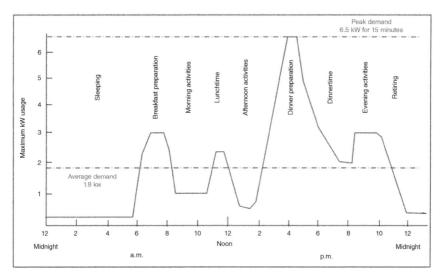

Figure 1.4 Load curves for a typical day. Source: US Department of Energy, Office of Electricity Delivery and Energy Reliability.

throughout the day. The peak demand occurs in the early evening when people return from work and are engaged in family activities or dinner preparation. The power demand rises and falls throughout the day depending on other activities, such as a peak when people are getting ready for work or troughs when they are sleeping. These load curves are constantly monitored and predicted by the utilities and operators to plan for the operation of the grid, and they are updated at regular intervals to account for changes in behavior, such as the COVID-19 pandemic.

The power distribution system is the last leg of the power delivery from the substations to the consumer. The three components of the power grid are usually defined by the voltage levels at which they operate at. Generation happens at generating stations at low voltages, following which the power is immediately transformed to much higher voltages on site. Generation plants send the power where they are stepped up till 20,000 V, following which they are fed to the transmission

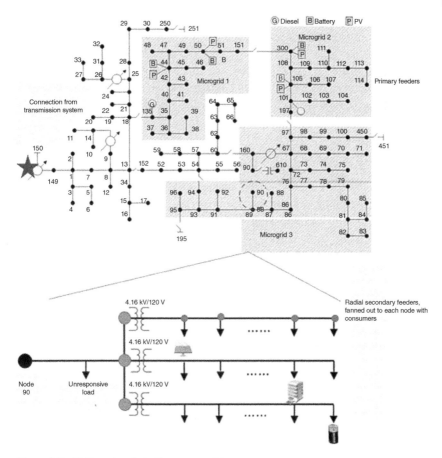

Figure 1.5 Voltage levels in the power grid.

grid where they can be stepped up as high as 765,000 V, commonly written and referred to as 765 kV. The power is stepped up to these very high values to reduce losses in transmission, which are directly proportional to the current and inversely proportional to the voltage. The distribution system substation is considered to be at the 13.2 kV level (or could be higher), following which the voltage is stepped down to be sent to the consumers. This structure is illustrated in Figure 1.5.

Energy control centers have traditionally been the decision centers for the electric generation and transmission centers. There are enabled by the wide area measurements fed to the control centers by the SCADA (Supervisory Control And Data Acquisition) and other measurement systems. The control center operator(s) is a key part of the overall operation of the grid with various responsibilities including but not limited to the following:

1. Monitor and react to key system performance indices such as voltage, frequency, power quality, and other metrics (such as reliability metrics).
2. Respond to emergencies and alerts – the control system operator has to handle the alerts from various algorithms and applications running at the control center. In addition, they also deal with emergencies such as trees hitting transmission lines or fires because of malfunctioning equipment.
3. Ensure system reliability by scheduling maintenance on equipment in anticipation of failures.
4. Respond to larger customer requests such as industries or other infrastructures. This could be a larger consumer who is testing their on-site back-up generation or infrastructural loads such as the transit system.
5. Coordinate with other stakeholders such as generation companies, transmission operators, utilities, and maintenance crews among others to ensure seamless operation.
6. Ensure that system operation is compliant with system regulations put in place by authorities such as FERC and NERC at all times.

In short, the control system is responsible for ensuring that electricity is being generated, transmitted, and distributed to the consumers in a safe and reliable manner. It coordinates all system operations with the other stakeholders by monitoring the performance and reacting to problems, ensuring that its operation is compliant with regulations at every instance.

1.1.1 Power Grid Operation

The power system is operated with support from a set of power "applications," which are monitor and control algorithms embedded into software tools based on the laws of physics. These algorithms allow the operator to understand the condition of the power grid at that moment and enable the operator to take decisions that can control the grid as desired. There are several power system applications that are critical to its operation, with several more being developed based on the new technologies that are being deployed in the grid. It is important to understand the key fundamental applications, as this will allow deeper understanding of the newer applications in the smart grid. Examples of these applications include the following:

1. Power flow
2. State estimation
3. Optimal power flow and economic dispatch
4. Continuation power flow
5. Automatic generation control (AGC)
6. Stability analysis

Power flow is the algorithm that determines the complex voltage at every node in a network, given the generator power injections and voltage set points, load-active and reactive power demands, and network impedances. The solution of the power flow is based on Kirchhoff's laws. There are multiple solution techniques for solving the power flow, of which the most common ones are (i) the Newton–Raphson technique, (ii) the Gauss–Seidel technique, and (iii) the fast decoupled load flow (FDLF). There are also techniques exclusive to distribution systems that take into account the radiality of the networks, which include techniques such as the forward backward sweep technique.

The Newton–Raphson method of solving the power flow requires the formation of a gradient matrix called the Jacobian. The Jacobian matrix is formed by eliminating the slack bus (which is the reference bus from which the voltage angles are determined) and the voltage buses from the bus data. The Jacobian matrix is actually a combination of four different matrices: The real power (P) and reactive power (Q) are differentiated with respect to voltage (V) and angle (θ). The updated voltage and angles are found out by multiplying the inverse of the Jacobian with the changes in P and Q. The above process is iterated until the difference in the mismatch vectors (ΔP and ΔQ) for successive iterations are small enough than the tolerance value. In this case, the tolerance value is set at 0.01. This indicates that the power flow solution is accurate up to 0.01 pu, which is the accepted tolerance condition. The fast decoupled power flow technique is similar to the Newton–Raphson method, except that the Jacobian remains constant. This is achieved through two assumptions:

1. The resistance $R = 0$
2. Difference in angle $\theta_i - \theta_j = 0$

Together, these two assumptions indicate that J11 and J22 (the diagonal elements of the Jacobian) become constant and is equal to the imaginary part of the Y Bus. Hence, for the FDLF method, the Jacobian remains constant throughout the iterating process. Like the Newton–Raphson method, the FDLF method also looks at the mismatch vector to find out when to stop iterating. Because of the assumptions made in the solution, the FDLF method takes more iterations to converge but is faster to compute as the Jacobian remains unchanged.

Continuation power flow is used to determine the stability of the system. In the continuation power flow, the power flow is solved continuously by changing the load conditions for each time. This is not possible in the usual power flow as the Jacobian becomes singular after some time. To solve this problem, a predictor–corrector technique is used.

State estimation is used to find the condition of the system. The usual measurements used are voltage (V), real (P), and reactive (Q) powers and line flows $(P_{ij}, P_{ji}, Q_{ij}, Q_{ji},)$ from both ends of the line. The combined matrix of all these measurements is referred to as the "Z" matrix. Some errors are usually assumed to be present in these measurements, from equipment malfunctions or loss of data in communication. State estimation can be performed through various techniques, but the most popular method is the weighted least squares (WLS) technique, which allows weighing of measurements to account for inaccurate measurements. Another important part of state estimation is the detection of bad data in the measurements, and the most common method for this is the chi-squared statistical test. This test looks at the probability of a value lying outside a given range for the degree of freedom and the accuracy.

These power system applications together enable the operator to have situational awareness on the state of the grid and enable them to take control actions to mitigate any problems. In the context of the smart grid, it is important to note that the timeline for the power system applications are becoming shorter because of the increasing application requirements.

1.2 What Can Go Wrong in Power Grid Operation

The power grid infrastructure is vast, spread across a wide geographical area, and consists of various components. Considering the scale of the power grid, it is exposed to various threats and has several vulnerabilities. While the power grid is designed with multiple redundancies to operate through various contingencies, it can still suffer from failures. These failures are often a result of power grid elements not performing as expected or because of external disturbances. Failures can occur in individual components such as generators, transmission lines, and measuring and monitoring equipment or may happen across multiple components. In general, when these failures occur at the transmission level, they are termed as "faults." There can be a variety of faults in the grid caused by external or internal sources:

1. Lightning
2. Wind and snow
3. Deterioration of materials (insulation, conductors, etc.)
4. Trees
5. Motor vehicle accidents (such as people driving into electric poles)
6. Human errors (mistakes in interpreting situations and wrong control actions)
7. Cyber attacks
8. Animals (mainly squirrels)

Faults are dangerous situations as they are largely uncontrolled and can lead to dangerous situations such as arc flashes. These can be fatal to people and cause expensive damage to equipment, which often takes a long time to repair.

Hence, the power grid has various levels of "protection" built into it as a first line of defense. These protective systems try to isolate the faults from the rest of the power grid to (i) stop the flow of energy to the affected area and (ii) prevent the fault from cascading to other parts of the grid. Protection equipment are mainly either (i) fuses or (ii) circuit breakers.

Fuses are used mostly in distribution circuits. They are capable of both detecting and isolating faults, but with less flexibility. They are specially designed and rated wires that melt when the current passing through them is higher than a specified tolerance. However, fuses are very simple to install and are inexpensive, but a problem with fuses is that they need to be replaced in case they are "activated," as they simply melt to break the electrical connection. For this reason, fuses are more prevalent in the distribution circuits where outages affect a smaller number of people.

Circuit breakers and switches come in different types, but a major difference from fuses is that they are re-usable and can sometimes be operated remotely. Circuit breakers operate in tandem with a relay, which performs the sensing and monitoring functions. The relay is responsible for detecting the presence of a fault by monitoring the current and other factors and send a "trip" signal to the circuit breaker, which then breaks the electrical connection by "opening" the circuit. Relays are often programmed or "set" by a variety of complex mechanisms for detecting a wide variety of faults. These are referred to as relay protection "schemes," whose examples include distance protection scheme, undervoltage or overvoltage protection, and generator protection. Circuit breakers can be mechanical or solid-state devices. They are also deployed at the low-voltage level, most commonly in breaker panels in houses. However, these low-voltage devices are different from the high-voltage devices, although both work by isolating the fault from the rest of the circuit when it detects a fault. In the smart grid, smart circuit breakers are becoming more ubiquitous, especially at places where the circuit breakers are often tripped. Smart circuit breakers enable the operator to operate it from the control center remotely, thus significantly reducing the time it takes to energize a circuit after a fault, as compared to sending out a maintenance truck to turn it on again.

1.3 Learning from Past Events

The power grid infrastructure is unique from other critical resources such as water in that the power grid cannot "store" electricity anywhere in the system in bulk. Although batteries are capable of storing electricity, they cannot hold it for a long duration, and the process of storing and retrieving electricity from batteries inherently comes with losses. Moreover, having large-scale batteries is still an expensive proposition for the electric grid. This lack of storage suggests that when failures occur in the power grid, there is very limited time to respond to such failures.

Electricity being utilized by the consumer is being generated at every instant, and there cannot be a pause in the generation process. When a fault occurs in the grid, the rest of the generating resources need to increase their production immediately to ensure that the shortfall is covered. Because the generators are usually huge rotating machines that generate AC power, they provide the grid with an inherent inertia that takes care of the shortages for a few cycles. In the meantime, because the grid works synchronously at the same frequency, other generators that are connected try to spin faster when a loss of a generator leads to a drop in the frequency. However, when the shortage is not mitigated quickly either through automated responses or through operator control actions, the generators fall out of synchronism, leading to a loss of electricity transmission and consumers being disconnected from the grid. This situation is called as a "blackout." Table 1.1 shows a list of well-known blackouts that have resulted in a large number of people being deprived of power. Blackouts lead to life-threatening situations with tremendous economic consequences.

Blackouts often prove to be watershed moments in grid operations. For example, the 1965 blackout in the NE United States leads to the development of the modern

Table 1.1 List of well-known blackouts around the world.

Location	Date (MM/DD/YY)	Scale (MW or population)	Blackout time
US – Northeast	10–11/9/65	20,000 MW, 30 M	13 minutes
New York	7/13/77	6,000 MW, 9 M	1 hour
France	1978	29,000 MW	26 minutes
Japan	1987	8,200 MW	20 minutes
US – West	07/02/96	11,700 MW	36 seconds
US – West	8/10/96	30,500 MW	>6 minutes
Brazil	3/11/99	25,000 MW	30 seconds
US – NE	8/14/03	62,000 MW, 50 M	>2 hour
London	8/28/03	724 MW, 476 K	8 seconds
Denmark and Sweden	9/23/03	6,500 MW, 4 M	7 minutes
Italy	9/28/03	27,700 MW, 57 M	27 minutes
India	7/30/12	48,000 MW, 600 M	>6 hours
Brazil	03/21/18	18,000 MW, 10 M	>4 hours
Ukraine	12/23/15	73 MW, 230 K	6 hours
US – Texas	2/2021	15,000–20,000 MW	hours–days (rolling)

control centers, with wide area measurements and better situational awareness and control. Similarly, the 2003 blackout is also a significant event in the operation of the power grid, especially in the United States. The August 2003 blackout is one of the largest in the history and shut down 263 power plants (531 units) in the United States and Canada. It left over 50 million people without power at various times and affected eight states in the United States (Michigan, Ohio, Pennsylvania, New York, Vermont, Massachusetts, Connecticut, and New Jersey) and the province of Ontario in Canada. Approximately 62,000 MW of load was lost, and power was restored as early as 2 hours into the blackout, while other people were left without power for up to 14 days. A simplified sequence of events is described as follows:

1. 1 : 30 – Loss of East Lake generator because of over excitation
2. 2 : 02 – Loss of Stuart–Atlanta transmission line because of tree contact
3. 2 : 02 – Midwestern ISO's system model becomes inaccurate, not reflecting the changes in the field
4. 2 : 14–3 : 08 – Software bug in FirstEnergy's control center, leading to continuous alarms for over an hour that went unnoticed
5. 3 : 05 – Loss of Harding–Chamberlain transmission line because of tree contact
6. 3 : 32 – Loss of Hanna–Juniper transmission line because of tree contact
7. 3 : 41 – Loss of Star–S. Canton transmission line because of tree contact
8. 4 : 06 – Loss of Sammis–Star transmission line because of incorrect protection system operation, where a high overload on the line was wrongly detected as a fault in Zone 3 of the protection system

Because of an initial set of faults that went undetected, oscillations in voltage and current and very high currents due to high transmission line loads lead to many transmission lines being tripped. They were sometimes falsely tripped because of the miscoordination in the protection system settings, which lead to many false Zone 3 trips, where very high currents are misinterpreted as fault currents. As a few generators started tripping, the load generation imbalance caused underfrequency in the system and lower voltages. The generators then started tripping because of this underfrequency, overexcitation, and out-of-step conditions, and the generators started losing synchronism with the grid.

The 2003 blackouts lead to a major inquest into power system operations and lead to significant changes. A committee investigated the incident and came up with several main causes of the blackout including the following:

1. There was inadequate situational awareness at FirstEnergy (FE). FE did not recognize/understand the deteriorating condition of its system.
2. FE failed to adequately manage tree growth in its transmission rights of way.
3. Failure of the interconnected grid's reliability organizations (mainly MISO) to provide effective real-time diagnostic support.

4. FE and East Central Area Reliability (ECAR) failed to assess and understand the inadequacies of FE's system, particularly with respect to voltage instability and the vulnerability of the Cleveland–Akron area, and FE did not operate its system with appropriate voltage criteria.
5. No long-term planning studies with multiple contingencies or extreme conditions had been performed
6. No voltage analysis was performed for the Ohio area, and inappropriate operational voltage criteria were used
7. No independent review or analysis of FE's voltage criteria and operating needs was performed
8. Some of NERC's planning and operational requirements were ambiguous

The blackout led to the federal government of the United States to include reliability provisions in its Energy Policy Act of 2005. NERC's operational guidelines that were not enforceable became mandatory requirements for US electricity providers to comply with. This led to widespread changes in their operation and higher awareness among public and officials on the ease with which the grid could be brought down. The development of a smart grid can in some ways be considered as one of the effects of the 2003 blackout.

1.4 Toward a Smarter Electric Grid

To mitigate potential failures, a key element is to obtain better situational awareness so that operators can respond to contingencies with up-to-date information. The smart grid is a combination of the electric grid with other enabling infrastructures. The cyber-physical model of the smart grid consists of the following layers as shown in Figure 1.6: (i) physical power system layer, (ii) device/sensor layer (responsible for measurement and control), (iii) communication/cyber layer, and (iv) management/application layer. The cyber-physical system architecture differs

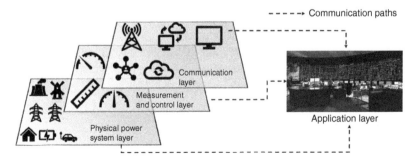

Figure 1.6 Cyber-physical system layers. Source: JOHN WILEY & SONS, INC.

from the OSI/network layers in that these layers are across various domains and are not used to describe the communication between various components.

The physical layer consists of the power infrastructure of the microgrid. This includes components such as generators, transformers, loads, and circuit breakers. There are well-defined physical models for all these components, which help in monitoring the performance of the system accurately.

The sensor and actuator layers consist of the devices that are responsible for sensing information about the state of the system and implement control decisions. Measurement devices sense various parameters such as voltage, current, frequency, and circuit breakers' status. Control devices include generator controllers, relays to trip the circuit breakers, and other local control devices.

The communication layer consists of the ICT infrastructure and is responsible for delivering information to relevant layers. This layer usually consists of devices such as switches, routers, and the communication medium. The communication can be wired or wireless depending on the requirements.

The application layer is central to monitoring and control that is responsible for the operation of the microgrid. This application receives data from the measurement layers through the ICT infrastructure and decides on a control action to be executed, if necessary. The control action is then sent back to the control layer through the ICT infrastructure. The cyber-physical smart grid is the way forward for the power grid, with increasing integration of digital devices and advanced communication and computation in the grid. It has and will continue to fundamentally alter the way the electric power grid operates. For the future workforce, it is important to not only have understanding of the physics of the power grid but also gain understanding in the advanced communication, computation, and security aspects of the power grid.

1.5 Summary

In this chapter, we have studied the basics of the electric power grid, its underlying physics, and operating structure. Failures and faults in the power grid were explained, along with a study of the most well-known blackouts in the grid. These blackouts and how they have motivated the transition to the smart grid were examined.

1.6 Problems

1 What is the highest AC transmission line voltage in the United States?
   A 345 kV

B 765 kV
C 500 kV
D 450 kV
E 1000 kV

2 Which of the statement is NOT correct about electric grid protection:
   A Circuit breaker detects the fault
   B Relay detects the fault
   C Fuse detects the fault
   D Circuit breakers open the line
   E Relay operates the breaker

3 Which of the statement is NOT correct about the existing electric grid:
   A The average age of the distribution transformer is greater than 50 years
   B A relatively high percentage of workforce is going to retire in next 5 years
   C The electric grid has one of the lowest R&D expenditures compared to other infrastructures
   D The distribution system is more visible than the transmission system for situational awareness
   E Operation is based on load following and load prediction

4 What are the major causes of blackout?
   A Loss of generation
   B Incorrect setting of protection systems
   C Squirrels
   D Human error
   E All of the above

5 Which of these is NOT usually considered as one of the smart grid layers?
   A Physical layer
   B Device/sensor layer
   C Cybersecurity layer
   D Management/application layer

1.7 Questions

(1) Explain the flow of electricity from generation to a consumer?
(2) What are the major power system applications?
(3) What are the main differences between a fuse and a circuit breaker?
(4) What were some of the major reasons the 2003 US Northeast blackout?
(5) Explain the different layers of a smart grid?

Further Reading

Amin, S.M. and Wollenberg, B.F. (2005). Toward a smart grid: power delivery for the 21st century. *IEEE Power and Energy Magazine* 3 (5): 34–41. https://doi.org/10.1109/MPAE.2005.1507024.

Andersson, G., Donalek, P., Farmer, R. et al. (2005). Causes of the 2003 major grid blackouts in North America and Europe, and recommended means to improve system dynamic performance. *IEEE Transactions on Power Systems* 20 (4): 1922–1928. https://doi.org/10.1109/TPWRS.2005.857942.

Fang, X., Misra, S., Xue, G., and Yang, D. (2012). Smart grid — the new and improved power grid: a survey. *IEEE Communication Surveys and Tutorials* 14 (4): 944–980. https://doi.org/10.1109/SURV.2011.101911.00087.

Farhangi, H. (2010). The path of the smart grid. *IEEE Power and Energy Magazine* 8 (1): 18–28. https://doi.org/10.1109/MPE.2009.934876.

FERC Staff (2020). Energy Primer: A Handbook of Energy Market Basics. https://www.ferc.gov/sites/default/files/2020-06/energy-primer-2020_Final.pdf (accessed 27 August 2022).

Pourbeik, P., Kundur, P.S., and Taylor, C.W. (2006). The anatomy of a power grid blackout - root causes and dynamics of recent major blackouts. *IEEE Power and Energy Magazine* 4 (5): 22–29. https://doi.org/10.1109/MPAE.2006.1687814.

2

Sense, Communicate, Compute, and Control in a Secure Way

In Chapter 1, we studied the power grid infrastructure and the various layers of a cyber-physical smart grid. The power grid infrastructure does not exist, in vacuum, as the cyber-physical nature of the grid dictates that various supporting infrastructures are needed to enable safe and secure operation of the power grid. The main objectives of the power grid can thus be considered to be as follows:

1. Reliable supply of power to customers
2. Secure and economic operation under several constraints

For a reliable, economic, and secure operation of the smart grid, it is important to coordinate with different components of the smart grid. The power grid operators need to be able to predict the system status at a given time to understand its operation. The power grid has been around for about 100 years, and hence, the operators are intimately familiar with its operation and are able to predict fairly accurately the system status. However, the power grid is vast, and the system status is constantly changing because of the stochastic nature of its components. Hence, it is important to deploy sensors that measure the system status accurately.

The first power "grid" was demonstrated in Manhattan, NY, USA, and alternating current (AC) technology won the race as the choice of power supply against direct current (DC) because of its suitability for transferring over long distances. Since those times, the power grid has grown to be a vast engineering marvel spanning the entire country and similarly across different continents in the world. While sensing is important to ensure that the operators know the system status, it is insufficient to sense just local information. Because of the connected nature of the grid, it is important for the operators to understand the status of various components at different locations and control devices at other locations. Hence, communication is essential to enable the smart grid.

While the early grids could be controlled from a single location using analog controls, the smart grid provides a lot of data that the operator needs to understand and visualize, based on which control actions need to be performed. Hence, digital

Cyber Infrastructure for the Smart Electric Grid, First Edition.
Anurag K. Srivastava, Venkatesh Venkataramanan, and Carl Hauser.
© 2023 John Wiley & Sons Ltd. Published 2023 by John Wiley & Sons Ltd.

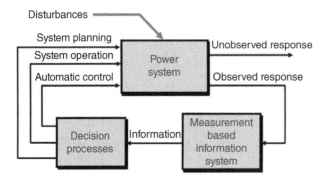

Figure 2.1 Sensing, communication, computation, and control for the cyber-physical smart grid.

computation and control becomes important for efficient operation. To summarize, for safe and efficient operation of the power grid, the following functions become important (Figure 2.1):

1. Sensing
2. Communication
3. Computation
4. Control

2.1 Sensing in Smart Grid

The measurements used in the smart grid for the physical power grid can be classified into continuous and discrete measurements. The continuous measurements are not truly "continuous" but rather sampled at a high frequency. Continuous measurements include current and voltage measurements at various points, power injection and consumption measurements that are sometimes computed rather than measured, and frequency measurements. Discrete measurements include switch statuses, fault indicators, and so on. Newer sensors include smart meters, phasor measurement units (PMUs), and more.

The continuous measurements use physical sensors such as potential transformers and current transformers. These sensors have several problems associated with them. These include the following:

1. Noise in measurement
2. A variety of transducers used for different functions, which lead to different response rates
3. No direct time synchronization is possible for measurements from different locations

4. Different requirements for communication bandwidth/response request rates, which make it difficult to synchronize measurements for operation
5. Equipment is usually not modular, repairs and upgrades become very expensive
6. The output is usually a root-mean-square (RMS) value, instead of the higher resolution phasors

This does not mean that the analog measurements are useless; rather, they need to be supplemented by modern digital resources to ensure that the grid operates in an efficient manner.

2.1.1 Phase Measurement Unit (PMU)

A phase measurement unit (PMU) is a device that provides as a minimum synchrophasor and frequency measurements for one or more 'three-phase AC voltage and/or current' waveforms. The synchrophasor and frequency values must meet the general definition and minimum accuracy required in the IEEE Synchrophasor Standard, C37.118-2011. The device must provide a real-time data output that conforms to C37.118.1 requirements. The measured values sampled generally every 4 seconds are displayed on the energy management system (EMS) screens. V and P, Q are generally transmitted via modems, microwave, or Internet directly to control rooms (Figure 2.2).

2.1.1.1 Why Do We Need PMUs?

Data from different locations are not captured at precisely the same time. While V, P, and Q normally do not change abruptly, there is a possibility of this happening if there is a large disturbance nearby. System monitoring becomes more critical during such disturbances, which results in a dynamic and transient behavior of the system. Faster synchronized data across different locations are needed to capture these dynamics. Fast real-time control to mitigate such dynamics is

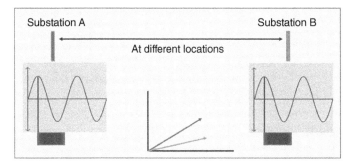

Figure 2.2 Capturing data from different locations synchronously.

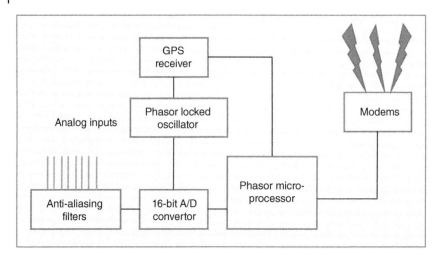

Figure 2.3 Phasor measurement unit architecture.

possible only with real-time situational awareness, which is enabled by synchronized, high-resolution measurement data.

However, what does "synchronized" measurements mean? The PMU is an estimation process that reconstructs the analog waveform using high-frequency sampling. The architecture of a PMU is similar to a digital relay or a digital fault recorder but with an added GPS receiver that ensures that the measurements are synchronized across all locations. A basic block diagram for a PMU is shown in Figure 2.3.

The PMU has an anti-aliasing filter, which restricts the bandwidth over which the analog waveform is sampled to reconstruct the waveform properly. There is a GPS receiver, which receives the synchronization signal, a phase-locked loop (PLL) oscillator, and an analog/digital (A/D) converter that all feed into an embedded microprocessor. The microprocessor is responsible for receiving these inputs and then estimating the phasor, which becomes the PMU. There are multiple phasor estimation algorithms, and they are usually proprietary to various PMU manufacturers. However, open-source PMU estimation algorithms such as openPMU are also available. Standards such as the IEEE C.37.118 control the performance standards for PMU, and the various estimation algorithms and the PMU themselves need to abide by these standards (Figure 2.4).

PMUs can not only measure current and voltage phasors but also provide other additional information. This may include frequency, rate of change of frequency (ROCOF), circuit breaker switch status, and so on. Using such synchronized measurements for monitoring the grid results in several advantages, the primary being that even when the measurements are taken at completely different geographical

Figure 2.4 Sampling rate of waveforms for PMUs.

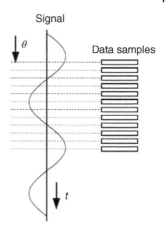

locations, say hundreds of miles apart, then they can still be put together on the same reference frame and compared to derive meaningful information from the data. This is demonstrated in Figure 2.2.

2.1.1.2 Estimation of Phasors

The "measurement" part of a PMU can be considered a misnomer, as the PMU is an estimation procedure based on signal processing. Most PMUs also estimate the voltage and current phasors only for the positive sequence quantity. Although theoretically a PMU can get individual data points from the different phases, say phases A, B, and C, to estimate the positive sequence phasor, this approach is prone to measurement noise.

Most phasor calculation in commercial PMUs uses a 1–4-cycle window, likely centering in that window. To reduce noise, some manufacturers use the average value over an even number of windows (2 or maybe 4). There is latency in the PMU itself based on the number of cycles and processing time. Various manufacturers use different procedures and algorithms to estimate their phasors. Using the PMU from the same manufacturer at least provides consistency of the phasor algorithm.

The original phasors present in the grid is shown in the left of Figure 2.5, and the positive sequence component isolated is shown in the right graph. The way that the positive sequence component of the symmetrical components is isolated is given by the following equation:

$$V_{abc} = \begin{bmatrix} V_0 \\ V_0 \\ V_0 \end{bmatrix} + \begin{bmatrix} V_1 \\ \alpha^2 V_1 \\ \alpha V_1 \end{bmatrix} + \begin{bmatrix} V_2 \\ \alpha V_2 \\ \alpha^2 V_2 \end{bmatrix} = \begin{bmatrix} 1 & 1 & 1 \\ 1 & \alpha^2 & \alpha \\ 1 & \alpha & \alpha^2 \end{bmatrix} \begin{bmatrix} V_0 \\ V_1 \\ V_2 \end{bmatrix} \tag{2.1}$$

In the above equation, V_a, V_b, and V_c are the original phasors, and V_a, 1, V_b, 1, and V_c, 1 represent their positive sequence components.

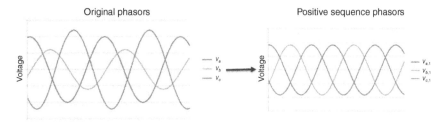

Original phasors   Positive sequence phasors

Figure 2.5 Original phasors and positive sequence components of phasors.

2.1.1.3 Phasor Calculation

Discrete Fourier transform (DFT) is used to estimate the phasors. Fourier coefficients from cosine and sine waves (as the waveforms we are trying to estimate are sinusoidal in nature) are multiplied with the samples from the waveform x_k, and the summation of the real and imaginary components are calculated. This gives us the phasor for the given waveform. Consider a signal X that can be represented in Cartesian coordinates as $X_r - jX_i$. Then, the phasors are given as follows:

$$X_r = \frac{\sqrt{2}}{N} \sum x_k cosk\phi$$

$$X_i = \frac{\sqrt{2}}{N} \sum x_k sink\phi$$

An important consideration in the estimation of phasors is that we cannot measure an instantaneous phasor, as the waveform needs to be observed over a specific measurement interval. The distinction here is that while the phasor value itself is instantaneous, it is estimated over a time interval of $k, k+1, k+2, \ldots$ depending on the number of observations required to estimate the phasor. This is illustrated in Figure 2.6.

2.1.1.4 Time Signal for Synchronization

The value of the PMU is tied with its ability to synchronize with a global reference. To achieve a common timing reference for the PMU acquisition process, it is essential to have a source of accurate timing signals (i.e. synchronizing source) that may be internal or external to the PMU. Hence, the time signal used for synchronization

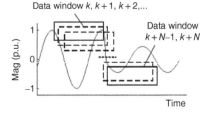

Data window $k, k+1, k+2,\ldots$

Data window $k+N-1, k+N$

Figure 2.6 Phasor estimation using DFT.

is regulated by standards. The synchronization signal is usually a GPS (Global Positioning System) signal, with 24 satellites (originally) in the medium Earth orbit. This system provides a 1–2 μs accuracy for the time signal. The GPS signal is usually given to the PMU in the form of IRIG-B pulses, or the IEEE 1588 Precision Time Protocol (PTP). The GPS signal is received by the PMU once every second, with the specified accuracy, and is then used to time stamp and synchronize the measurements.

For internal, the synchronization source is integrated (built-in) into the PMU (external GPS antenna still required). In the latter case, the timing signal is provided to the PMU by means of an external source, which may be local or global, and a distribution infrastructure (based on broadcast or direct connections). Within a PMU, a phase-locked oscillator is used to generate the time tags within the second. The time tag is sent out with the phasors. Thus, if a phasor information packet arrives out of order to a PDC (phasor data concentrator), the phasor time response can still be assembled correctly. If the GPS pulse is not received for a while, the time tagging error may result in significant phase error.

2.1.1.5 PMU Data Packets
In addition to standardizing the estimation and accuracy of PMUs, the IEEE C37.118 standard also lays down guidelines for the data packet used to communicate the PMU measurement between the field device and the PDC at the control center. The data packet structure for the PMU packet is shown in Figure 2.7. Synchrophasor measurements are tagged with the UTC time corresponding to the time of estimation. The data packet consists of three fields: a second-of-century (SOC) count, a fraction-of-second (FRACSEC) count, and a message time quality flag. The SOC count is a count of seconds from UTC midnight (00 : 00 : 00) of 1 January 1970, to the current second.

2.1.1.6 PMU Applications
PMUs present several advantages over traditional SCADA measurements (Figure 2.8):

1. Global behavior of the system may be inferred from local PMU measurement
2. Phasor measurement data can be used to supplement/enhance the existing control center functions and provide new functionalities

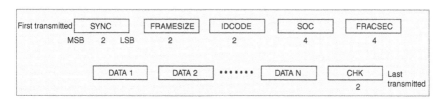

Figure 2.7 PMU data packet.

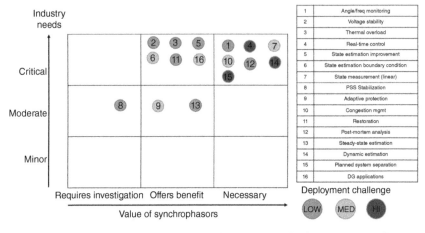

Figure 2.8 PMU-based applications for the future smart grid. Source: Adapted from Novosel and Vu [2006].

3. Phasor measurement data with GPS signal can provide synchronized voltage and current phasor measurements across a wide region
4. By measuring the phase directly, the power transfer between buses can be computed directly
5. High sampling rate (30 samples per second) that leads to higher resolution of data
6. Extended visibility beyond one's own operating region is provided by the synchronous nature of the measurement and the higher resolution
7. Disturbance monitoring -transient and steady-state responses can be effectively monitored

There are several other PMU-based applications in development, which may eventually become standard practice in grid operation. These include system restoration, model validation applications, and setting up more reliable protection systems, among others.

2.1.2 Smart Meters

While PMUs represent the technological advances in the transmission system, advanced metering infrastructure (AMI), which is primarily enabled by smart meters, represents the advances for distribution systems. AMI is an integrated system of smart meters, communications networks, and data management systems that enables communication between load serving entities and their customers. The system provides a number of important functions that were not previously possible or had to be performed manually, such as the ability to automatically and remotely measure electricity use, connect and disconnect service, detect tampering, identify and isolate outages, and monitor voltage among others. Combined with improvements in customer-level technologies, such as Internet of Things (IoT) devices, smart appliances, and automation technologies, AMI also make it easier for utilities to offer time-based rate programs and other incentives that encourage customers to participate in grid management and provide grid services such as mitigating peak demand and manage energy consumption and costs. AMI deployment usually has three key components:

1. Smart meters installed at the customer's premised that typically collect electricity consumption data at regular time intervals (typically every 5, 15, 30, or 60 minutes).
2. New or upgraded communications networks to transmit the large volume of data recorded by the AMI infrastructure to the utilities database.
3. A meter data management system (MDMS) to store and process the AMI data and to integrate meter data with other grid applications or business process systems, including head-end systems, billing systems, customer information systems (CIS), geographic information systems (GIS), outage management systems (OMS), and distribution management systems (DMS). (Not all utilities use a MDMS.)

2.1.2.1 Communication Systems for Smart Meters

Smart meters require communications networks that are capable of delivering large amounts of streaming data in a reliable manner. These communication networks connect customer end smart meters to head-end systems owned by the utilities, which manage data communications between smart meters and other information systems, including MDMS, CIS, OMS, and DMS. The head-end system transmits and receives data, sends operational commands to smart meters, and (occasionally) stores the data to enable operational support, such as billing.

Most utilities use a variety of wired and wireless communications technologies to build the AMI infrastructure, considering how each technology fits with their operational goals, service area characteristics, and business process constraints. The wired technologies include fiber optic cable, digital subscriber line (DSL), power line communication (PLC), and even dial-up models in some cases. Wireless technologies based on radio frequency (RF) are becoming more popular, with options being mesh systems, cellular networks, and so forth. In the absence of standardization, utilities typically combine multiple approaches and integrating with both legacy and new systems involving multiple vendor products.

In addition, many utilities use common communications platforms to support multiple field devices, including smart meters, customer systems, and distribution automation (DA) equipment. For example, fiber networks and wireless radio networks may use one protocol for switch status monitoring and another for smart metering.

2.2 Communication Infrastructure in Smart Grid

The communication requirements for supporting smart grids can be classified into (i) physical network infrastructure to ensure connectivity and (ii) protocols to support market function – including communication and data coordination. The bulk energy grid is well supported by optical fibers connecting control centers to substations, and multiple protocols such as synchronous optical networking (SONET) enable fast and secure grid communications. At the sub-transmission/distribution level, wide area network (WAN) or LTE-based networks are used in conjunction with various protocols such as DNP3 (IEEE 1815, legacy) or IEC 61850 to measure and control the grid. Market functions such as submitting bids are typically performed over the Internet using various authentication mechanisms to ensure security. This structure provides both the physical infrastructure required to implement and the appropriate standardization of operating procedures to ensure the smooth functioning of the market.

Further, with the advent of the IoT, connectivity and communication with grid edge resources and loads is enabled without the need to build additional infrastructure, providing more visibility and control capabilities throughout the distribution grid. Various communication protocols such as ZigBee, Modbus, IEEE Std 802.15.4, and PLC standards allow for communication with AMI devices. Recently, the IEEE 2030.5 Smart Energy Profile (SEP) 2.0 standard, which provides a framework for monitoring and control of DER assets, has been gaining traction with grid operators and has been suggested as the standardized communication protocol for DER aggregation programs. CAISO outlines in their Common Smart Inverter Profile (CSIP) how SEP should be implemented to meet Rule 21, requiring DERs to have monitoring and reporting capabilities, and grid support functionalities such as Volt-VAr Control (VVC).

2.3 Computational Infrastructure and Control Requirements in Smart Grid

Control centers have evolved over the years into a complex communication, computation, and control system. Control centers of present day tend to have three separate components (Figure 2.9):

1. Supervisory control and data acquisition (SCADA) system

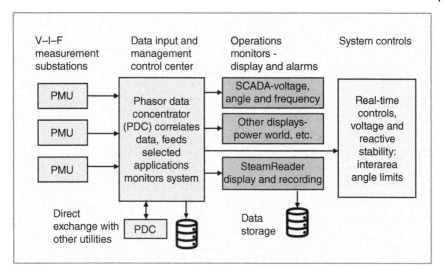

Figure 2.9 Communication, computation, and control for smart grid.

2. Energy management system (EMS)
3. Business management system (BMS)

A control center fulfills certain functions key to the operation of a power system. The implementations of these functions in a computational point of view are called applications.

The first group of functions is largely driven by the communication system and is related to data acquisition and the mechanism of supervisory control. Typically, data acquisition function collects real-time measurements of voltage, current, real power, reactive power, breaker status, transformer taps, etc., from substation RTUs every 2 seconds to get a snapshot of the power system in steady state. The collected data is stored in a real-time database for use by other applications. The sequence of events (SOE) recorder in an RTU is able to record more real-time data in finer granularity than they send out via the SCADA system. These data are used for possible post-disturbance analysis. Indeed, because of the SCADA system limitations, there are more data bottled up in substations that would be useful in control center operations. These issues are mitigated with advanced sensing systems such as PMUs, as discussed previously.

The second group of functions is for power system operation and is largely derived from the traditional EMS. They can be further classified under generation control and network (security) analysis and control. These applications include state estimation, security-constrained unit commitment (SCUC), security-constrained economic dispatch (SCED), contingency or security analysis, load management, automatic generation control (AGC), and others. It is important to note here that the term "security" here refers to network security or ability of the power system to supply energy to its loads within power quality constraints. Short-term load forecasts in 15 minutes intervals are carried out in the control center to drive these applications. State estimation is used to cleanse real-time data from SCADA and provide an accurate state of the system's current operation. A list of possible disturbances, or contingencies, such as generator or transmission line outages, is postulated, and against each of these contingencies, the power system's response is studied to ensure that the system's operational state is robust to these failures. This is called contingency analysis or security analysis. The objective of the contingency analysis is to ensure that the power system is at least $(n-1)$ stable, i.e. it is robust to at least one equipment failure at the transmission level. AGC is used to balance power generation and load demand instantaneously in the system, by redirecting power flows in the transmission lines. Unit commitment and economic dispatch are driven by the load forecasts and the BMS function, where market participants bid to provide grid services.

The third group of functions is for business applications and is the BMS. For an ISO/RTO control center, it includes market clearing price determination, congestion management, financial management, and information management. Different market structures exist, and some regions function with only bilateral contractual agreements, rather than a market structure. The three groups of functions and the interaction between them are shown in Figure 2.10.

2.3.1 Control Center Applications

State estimation is used to find the condition of the system. The usual measurements used are voltage (V), real (P) and reactive (Q) powers, and line flows ($P_{ij}, P_{ji}, Q_{ij}, Q_{ji}$) from both ends of the line. The combined matrix of all these measurements is referred to as the 'z' matrix. Some are usually present in these measurements. These errors usually result from sensing equipment malfunction such as CT/PT errors or from faulty communication such as missed/malformed packets. The state estimation problem can then be stated as the maximum likelihood estimation (MLE) that estimates the state $\hat{x}$, the solution that minimizes the measurement error $z - h(x)$. The estimates of the system state are then calculated using iterative methods, such as Newton–Raphson or Gauss–Seidel. The mismatch vector gives the change in voltage and angles, which is then updated, and the process is iterated until the mismatch vector becomes less than tolerance. The state estimation also uses a statistical method such as the chi-squared test to detect any bad data in the measurements. This test looks at the probability of a value lying outside a given range for the degree of freedom in the measurements.

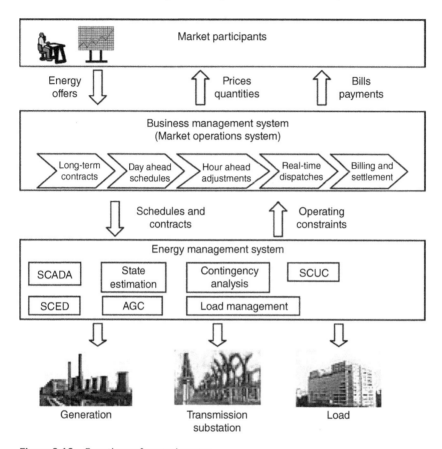

Figure 2.10 Functions of control center.

Continuation power flow is used to determine the stability of the system. In the continuation power flow, the power flow is solved continuously by changing the load conditions for each time. This is not possible in the usual power flow as the Jacobian becomes singular after some time. The purpose of continuation power flow is to determine the voltage collapse point of the system.

Unit commitment and economic dispatch: Unit commitment is the process of deciding when and which generating unit at each power station should start-up and shut-down to create an overall generation schedule to meet the load. Economic dispatch is the process of deciding what the individual power outputs should be of the scheduled generating units at each time period to ensure that the schedules are *economical*. Unit commitment and economic dispatch are usually constrained by the power flow to ensure that all of the load is supplied, which is referred to as security-constrained unit commitment (SCUC)

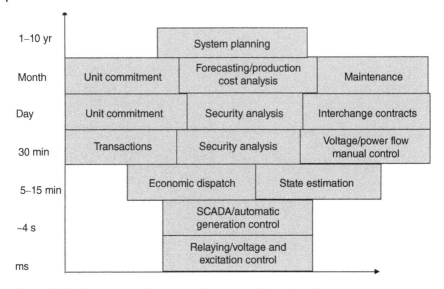

Figure 2.11 Various control center applications and their timelines.

and security-constrained economic dispatch (SCED). These are very challenging optimization problems because of the astronomically large number of possible combinations of the on and off states of all the generating units in the power system over all the time periods.

These control center applications all have different timelines, as illustrated in Figure 2.11.

2.4 Cybersecurity in Smart Grid

The deployment of a smart grid requires a deeper understanding of the interconnections between physical, communication, computation, and control layers and the potential impacts resulting from successful cyber attacks. While technologies discussed in this chapter such as PMU, wide area measurement systems, and AMI, will be deployed to help achieve better grid operations, they also present an increased dependency on cyber resources that leads to more vulnerabilities and a larger attack surface for malicious agents. The North American Electric Reliability Corporation (NERC) has recognized these concerns and introduced compliance requirements to enforce baseline cybersecurity throughout the bulk power system. Current events have shown the trend of attackers using increasingly sophisticated attacks against industrial control systems, and numerous countries have acknowledged that cyber attacks have targeted their critical infrastructures.

The underlying cyber-physical relationship of smart grid infrastructure can lead to unintended system dependencies.

An adversary could exploit vulnerabilities in the communication and computation infrastructure (also referred to as cyber infrastructure) and create attacks designed to either corrupt the content (i.e. integrity attacks), or manipulate the communication to make data unavailable (i.e. availability attacks such as denial of service [DoS], desynchronization, and timing attacks), or gain access to privileged information (i.e. confidentialty attacks) essential to the operation of the power grid. It is important to study and analyze the impacts of such attacks on the power system as they could severely affect its security and reliability. These impacts can be measured in terms of loss of load or violations in system operating frequency and other operational constraints. Detailed analysis of attack scenarios will also help develop countermeasures that can prevent attacks or mitigate the impact from attacks.

2.4.1 Methods to Provide Cybersecurity for Smart Grids

Defense techniques against cyber attacks can take the form of pure physical defense methods such as deploying additional equipment, or pure cyber defense techniques such as intrusion detection systems (IDS), firewalls, or access control policies, or cyber-physical defense techniques that work based on a combination of these approaches to quickly react to attacks on system operation. These techniques will be explored further in later chapters in the book.

2.5 Summary

This chapter dealt with sensing, communication, computation, and controlling the grid in a secure way for cyber-physical smart grids. Existing state-of-the-art methods have been detailed, and some ideas on the evolving smart grid infrastructure have been presented. These concepts are essential to understand the components of a cyber-physical smart grid and the interdependence of the various components associated with it.

2.6 Problems

1 Which of the statement is correct about analog and digital sensors in a smart grid:

   A Analog sensors work better than digital sensors in the presence of limited noise

B Analog sensors generally measure RMS values compared to phasor

C Digital sensors work better in the presence of very high noise level

D Analog sensors provide more flexibility compared to digital sensors

E None of the above

2 What does synchronizing measurements mean in a PMU?

A Voltage and current measurements are synchronized

B Network information (IT) and measurements (OT) are synchronized

C Data from different locations are synchronized using GPS

D Data from different utilities are synchronized

E Data from different vendors of measurement equipment are synchronized

3 Which one of these protocols is NOT a power system protocol?

A IEEE 2030.5

B IEEE C37.118

C IEEE 754-1985

D IEC 61850

E IEEE 1815

4 Which one of these can NOT be defined as the optimal power flow:

A Security-constrained power flow

B Constrained economic dispatch

C Power flow with minimum losses

D Transient stability

E None of the above

5 Which one of the following statements is INCORRECT about cybersecurity for the power grid?

A Recent technologies such as PMU, smart meters have been developed considering cybersecurity and can be considered safe

B Cybersecurity for the power grid needs to be a combination of physical and cyber measures

C Data availability issues such as Denial of Service (DoS) can affect power grid applications

D Impact of cyber attacks on the power grid can be measured by metrics such as loss of load

E There are mandatory compliance requirements that utilities have to follow to protect against cyber threats in the United States.

2.7 Questions

(1) Explain the various components of a phasor measurement unit (PMU) system.
(2) How are PMU measurements synchronized?
(3) Explain the communication architecture used for smart meters.
(4) List and discuss three power system applications.
(5) Explain the types of impacts that cyber attacks can have on the power grid and possible methods of mitigation.

Further Reading

De La Ree, J., Centeno, V., Thorp, J.S., and Phadke, A.G. (2010). Synchronized phasor measurement applications in power systems. *IEEE Transactions on Smart Grid* 1 (1): 20–27. https://doi.org/10.1109/TSG.2010.2044815.

Gungor, V.C., Sahin, D., Kocak, T. et al. (2011). Smart grid technologies: communication technologies and standards. *IEEE Transactions on Industrial Informatics* 7 (4): 529–539. https://doi.org/10.1109/TII.2011.2166794.

Gungor, V.C., Sahin, D., Kocak, T. et al. (2013). A survey on smart grid potential applications and communication requirements. *IEEE Transactions on Industrial Informatics* 9 (1): 28–42. https://doi.org/10.1109/TII.2012.2218253.

von Meier, A., Stewart, E., McEachern, A. et al. (2017). Precision micro-synchrophasors for distribution systems: a summary of applications. *IEEE Transactions on Smart Grid* 8 (6): 2926–2936. https://doi.org/10.1109/TSG.2017 .2720543.

Meliopoulos, A.P.S., Cokkinides, G., Huang, R. et al. (2011). Smart grid technologies for autonomous operation and control. *IEEE Transactions on Smart Grid* 2 (1): 1–10. https://doi.org/10.1109/TSG.2010.2091656.

Novosel, D. and Vu, K. (2006). Benefits of PMU technology for various applications. *Proceedings of the Seventh Symposium on the Management System of EES HK CIGRE*, Vol. 5, no. 8.11.

Sridhar, S., Hahn, A., and Govindarasu, M. (2012). Cyber–physical system security for the electric power grid. *Proceedings of the IEEE* 100 (1): 210–224. https://doi.org/10 .1109/JPROC.2011.2165269.

Werbos, P.J. (2011). Computational intelligence for the smart grid-history, challenges, and opportunities. *IEEE Computational Intelligence Magazine* 6 (3): 14–21. https:// doi.org/10.1109/MCI.2011.941587.

3

Smart Grid Operational Structure and Standards

This chapter deals with the operational structure and standards governing the smart grid, with a particular focus on the structure in the United States. A brief commentary on the operational structures from the rest of the world is included at the end of this chapter. Electricity generated at power plants moves from generation to transmission and distribution systems before it reaches the customers. In the United States, the power system consists of more than 7300 power plants, nearly 160,000 miles of high-voltage power lines, and millions of low-voltage power lines and distribution transformers, which connect 145 million customers, by the latest numbers from the US Energy Information Administration (EIA).

As we know, transmitting electricity over long distances leads to losses because of the resistance in the transmission lines. Hence, electric grids transport power at very high voltages to minimize losses, leading to interconnections between "local" installations. In the United States, there are three main such interconnections, which mainly operate independently from each other and only exchange limited power. These three interconnections are the (i) Eastern interconnection, (ii) Western interconnection, and (iii) Electric Reliability Council of Texas (ERCOT). Each interconnection has at least one Balancing Authority (BA), which is responsible for maintaining the electricity balance between demand and generation within its region. The BA does this by controlling the generation and transmission of electricity throughout its own region and coordinating with its neighboring Balancing Authorities.

The Eastern Interconnection encompasses the area from the east of the Rocky Mountains including a portion of northern Texas. The Eastern Interconnection consists of 36 balancing authorities of which 31 are in the United States and 5 are in Canada. The Western Interconnection encompasses the area from the west of the Rocky mountain to the coast and consists of 37 balancing authorities of which 34 are in the United States, 2 are in Canada, and 1 in Mexico. The ERCOT covers most of Texas (except regions covered by the Eastern Interconnection) and consists of a single BA. The interconnections help in making electricity more reliable

Cyber Infrastructure for the Smart Electric Grid, First Edition.
Anurag K. Srivastava, Venkatesh Venkataramanan, and Carl Hauser.
© 2023 John Wiley & Sons Ltd. Published 2023 by John Wiley & Sons Ltd.

by providing alternate and redundant routes for the supply of electricity from the generating stations to the consumers.

These interconnections describe the physical system of the grid. The actual operation of the electric system is managed by entities called balancing authorities. Most, but not all, balancing authorities are electric utilities that have taken on the balancing responsibilities for a specific portion of the power system. All of the regional transmission organizations (RTOs) in the United States also function as balancing authorities. ERCOT is unique in that the BA, interconnection, and the RTO are all the same entity and physical system.

While the interconnections represent the physical network structure of the grid, the operational structure is represented by the balancing authorities. The balancing authorities are typically electric utilities that have been assigned balancing responsibilities over a specific region assigned to them. In addition to the utilities, RTO and independent system operators (ISO) also function as balancing authorities (Figure 3.1).

A BA ensures that in real time, the power system demand and supply are finely balanced. This balance is needed to maintain the safe and reliable operation of the power system, failing which the frequency will deviate from the standard frequency at which the system operates, which is 60 Hz in the United States. The frequency deviation causes power quality deterioration, leading to damage to equipment and appliances, and can even cause local or wide-area blackouts.

Balancing authorities maintain this demand generation balance by ensuring that a sufficient supply of electricity is available to serve expected demand, which includes managing power flow with areas controlled by other balancing authorities. Balancing authorities are responsible for maintaining operating conditions under mandatory reliability standards issued by the North American Electric Reliability Corporation (NERC) and approved by the U.S. Federal Energy Regulatory Commission (FERC) and the Canadian regulators, if appropriate.

Figure 3.1 Role of balancing authority.

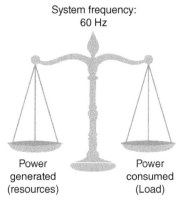

System frequency: 60 Hz

Power generated (resources)

Power consumed (Load)

Figure 3.2 Balancing authorities in the United States.

These operators monitor the grid to identify potential problems before a situation becomes critical. Figure 3.2 shows a map of the interconnections and balancing authorities in the United States.

3.1   Organization to Ensure System Reliability

FERC is the Federal Energy Regulatory Commission, a U.S. federal agency, and is an independent agency that regulates the interstate transmission of electricity, natural gas, and oil. In essence, FERC has a broad oversight into all aspects of the energy industry and as a federal entity has dominion over interstate commerce. On the other hand, NERC is the organization charged with creating reliability standards, enforcing and ensuring compliance with these standards, creating assessments, energy infrastructure security, ensuring workforce adequacy and training, and certification, among other responsibilities. FERC and NERC are complementary organizations and work together to achieve common goals. NERC is the electric reliability organization (ERO) certified by FERC to establish and enforce reliability standards for the bulk power system within the United States.

In September 2012, FERC announced the creation of the new Office of Energy Infrastructure Security (OEIS) that is charged with providing leadership, expertise, and assistance in identifying, communicating, and providing comprehensive solutions for any potential threats to energy facilities with the FERC jurisdiction from cyber attacks and related threats (according to ferc.gov). The formation of this office is an example of the evolving cyber-physical nature of the smart grid, and how regulatory standards for these evolving critical infrastructure assets are continuing to unfold in the United States.

NERC is cognizant of the interdependence in cyber-physical systems and has put in place a well-known and comprehensive cybersecurity program known as the NERC Critical Infrastructure Protection (CIP) standard. The NERC CIP program lays out general operating guidelines, while also specifying detailed operating requirements for ensuring the safety and security of critical infrastructure assets against cyber vulnerabilities. Other related efforts in this area include the Electricity Subsector Cybersecurity Capability Maturity Model ES-C2M2, which can be used to evaluate an organization's preparedness and resilience to cyber failures. While there are some steps toward standardization, more expansive and detailed sets of policies, procedures, and processes need to be in place.

NERC was founded in 1968 by representatives of various electric utilities, primarily for creating voluntary compliance standards required for the reliable operation of the bulk power system. NERC's current mission is to ensure reliability and security of the bulk power system, and NERC aims to do that not only by enforcing compliance with mandatory reliability standards but also by acting as a catalyst for utilities to exchange lessons learnt, helping utilities plan their resources and system operation with perspectives from other participants, and facilitating complaint behavior by identifying system weaknesses. In short, NERC

1. develops and enforces the operational standards for the bulk power system,
2. monitors the performance of the overall power grid,
3. leads planning studies to assess the adequacy of resources annually via a 10 year forecast and winter and summer forecasts,
4. audits asset owners, operators, and users for preparedness, and
5. educates and trains industry personnel.

3.1.1 Regional Entities

NERC works closely with eight regional reliability organizations (as shown in Figure 3.3) to fulfill its objectives. NERC oversees and coordinates with various entities from all segments of the electric industry: (i) federal agencies such as FERC; (ii) investor-owned utilities (IOUs); (iii) rural electric cooperatives; (iv) state, municipal, and provincial utilities that serve as load serving

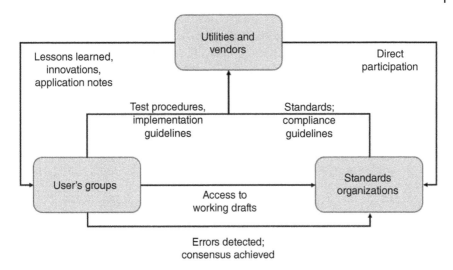

Figure 3.3 Interaction between utilities, enforcement agencies, and user groups.

entities; (v) independent power producers (IPPs); and (vi) end-use customers. Figure 3.3 shows an overview of how these entities work together, although the actual interactions happen across different levels.

3.2 Smart Grid Standards and Interoperability

Interoperability indicates the ability of complex and diverse systems, or components within such systems, to interact with each other and to operate effectively toward a common objective in an expected way without significant user intervention. Interoperability can be facilitated by the development of open standards, encouraging adherence to the standards or by enforcement of standards and compliance verification by the participants through independent audits and certification. Interoperability for the electric power system can be defined as "the seamless, end-to-end connectivity of hardware and software from the customers' appliances all the way through the distribution and transmission systems to the power source, enhancing the coordination of energy flows with real-time flows of information and analysis." Interoperability is a fundamental piece of the smart grid puzzle, as the evolving smart grid needs to be coordinated with resources from diverse domains and from various manufacturers and vendors.

In 2004, recognizing the advent of the smart grid with interdependencies between cyber and physical components, the U.S. Department of Energy (DOE) formed the GridWise Architecture Council (GWAC) to formulate interoperability concepts and facilitate the interoperation of smart grid technologies. Three years later, the Energy Independence and Security Act of 2007 (EISA) gave the National Institute of Standards and Technology (NIST) the primary responsibility to lead the development of a framework focusing on the protocols of operation and standards to achieve interoperability of smart grid devices and systems. The Recovery Act included $12 million grant specifically aimed toward developing a framework for smart grid interoperability that DOE transferred to NIST to fulfill its objectives under EISA.

In May 2009, NIST released the initial set of 16 interoperability standards, which addressed a wide range of subjects, including the roles of the utility, economics, smart meters, distributed generation components, and cybersecurity. NIST followed up on this initial effort with a report that included about 80 initial interoperability standards and 14 "priority action plans" to address gaps in the standards. The "NIST Framework and Roadmap for Smart Grid Interoperability Standards" was released in January 2010 and has been subsequently revised multiple times according to the current operating paradigms. The latest version of the release was version 4.0, released in February 2021.

In November 2009, NIST initiated the Smart Grid Interoperability Panel (SGIP), a new stakeholder forum to provide technical support to NIST's interoperability mission along with other stakeholders including Knoxville and EnerNex Corp, under a contract enabled by the American Recovery and Reinvestment Act. SGIP is currently focused on defining requirements for a smarter grid by driving interoperability, the use of standards, and collaborating across organizations to address gaps and issues hindering the deployment of smart grid technologies. The public–private partnership model of SGIP was established with the intent to ultimately transition to a non-profit, public/private funding model. This goal was successfully accomplished in 2013, and SGIP 2.0 was launched. Currently, SGIP is fully operational as a private entity that functions by creating and facilitating various working groups addressing different smart grid challenges such as the architecture group, the grid management group, the cybersecurity group, the distributed resources and generation group, and the testing and certification group. SGIP also coordinates closely with the GWAC and other standards setting organizations internationally.

3.3 Operational Structure in the Rest of the World

The rest of the world has evolving smart grid structures similar to the structure discussed for the United States. In Europe, for example, the Smart Grid Architecture Model (SGAM) framework has been proposed, which also references the NIST standards as its principles. Other countries from say Asia or South America tend to have a smaller number of regulatory authorities while still following the overall principles such as an emphasis on (i) cyber-physical system modeling and analysis, (ii) importance of interoperability, and (iii) regular audits and enforcement protocols, among others.

3.4 Summary

In this chapter, we examined the smart grid operational structure with a particular focus on the United States. We learned about the history of the organizations that develop and enforce operational standards, the various key stakeholders involved in the smart grid structure, and the challenges for both the participants and the regulatory agencies with the advent of the cyber-physical smart grid.

3.5 Problems

1 How many miles of high-voltage power lines are present in the United States approximately, according to the EIA?
   A 65,000 miles
   B 100,000 miles
   C 160,000 miles
   D 200,000 miles
   E 250,000 miles

2 Electricity is usually transmitted over long distances at very high voltages to -
   A Reduce losses caused by line resistance
   B Protect against electricity theft
   C Accommodate consumers who require energy to be delivered at 345 kV and above
   D Simplify the transmission, as electricity is generated at 345 kV
   E All the above

3 _____ is responsible for ensuring bulk power system reliability by enforcing the reliability standards.
   A Federal Energy Regulatory Commission (FERC)
   B US Government Department of Energy
   C Regional Transmission Operators (RTO)
   D National Institute of Standards and Technology (NIST)
   E North American Electric Reliability Corporation (NERC)

4 What are the security standards that entities that directly impact the reliability of bulk energy systems (including power producers, operators, etc.) are required to be compliant with?

A NERC Critical Infrastructure Protection (CIP) standards

B ISO 27001 set of standards

C NIST Cybersecurity Framework (CSF)

D DOE-STD-1192-2021, Security Risk Management

E None of the above

5 Which organization was primarily responsible for creating the interoperability standards as mandated by the Energy Independence and Security Act of 2007?

A Federal Energy Regulatory Commission (FERC)

B US Government Department of Energy

C Regional Transmission Operators (RTO)

D National Institute of Standards and Technology (NIST)

E North American Electric Reliability Corporation (NERC)

3.6 Questions

(1) Briefly explain the concept of interconnection in the North American electric grid and describe their areas.
(2) Explain the role of a balancing authority.
(3) What is the difference between FERC and NERC? Briefly explain the difference in their jurisdictions.
(4) Why is interoperability important for the electric grid? Explain which entity is primarily responsible for developing interoperability standards and your understanding of the key requirements from the standards.

Further Reading

Dolezilek, D. and Hussey, L. (2011). Requirements or recommendations? Sorting out NERC CIP, NIST, and DOE cybersecurity. *2011 64th Annual Conference for Protective Relay Engineers*, pp. 328–333. https://doi.org/10.1109/CPRE.2011 .6035634.

FERC (2022). About FERC: An Overview. https://www.ferc.gov/what-ferc (accessed 29 August 2022).

Mertz, M. (2008). NERC CIP compliance: we've identified our critical assets, now what? *2008 IEEE Power and Energy Society General Meeting – Conversion and*

Delivery of Electrical Energy in the 21st Century, pp. 1–2. https://doi.org/10.1109/
PES.2008.4596272.

NERC Staff (2012). The History of the North American Electric Reliability
Corporation. https://www.nerc.com/AboutNERC/Resource%20Documents/
NERCHistoryBook.pdf (accessed 29 August 2022).

4

Communication Performance and Factors that Affect It

4.1 Introduction

As has been described in the earlier chapters, smart grid ideas have been increasingly deployed during the early years of the twenty-first century. The smart grid relies on computation and communication to impart resilience and reliability to grid operations. In the next three chapters, we will describe in some detail the communication characteristics that affect the overall performance of smart grid systems and introduce the communication protocols that they rely on. The basis of the discussion is the fact that communication necessarily takes place over channels that require finite time to transmit signals and to propagate them. Furthermore, the channels are subject to various kinds of errors that can cause messages to be corrupted or lost. Building reliable systems in the face of such behaviors is challenging, but fortunately, these are problems that have been well researched over the years, and there are techniques that can overcome most of them.

Before proceeding to a detailed discussion of modern communication systems, it is essential that students become familiar with several concepts such as transmission rate, propagation speed, and error rate and that they understand how to use these quantities in calculations in order to predict or understand system performance.

The purpose of a communication system is to deliver data from a sender to a receiver. Data are represented as binary values, 1 or 0, known as *bits*. The number of bits per second that are sent is the *transmission rate*. *Messages* are usually made up of many bits, and when the message length and transmission rate are known, the *transmission time* for the message can be calculated. The formula is $T_{transmission} = L/R$, where T is the transmission time in seconds, L is the message length in bits, and R is the transmission rate in bits per second.

Scaling factors are often encountered in these calculations, and it is important to keep a track of them and adjust the formula to take them into account. First, as

Cyber Infrastructure for the Smart Electric Grid, First Edition.
Anurag K. Srivastava, Venkatesh Venkataramanan, and Carl Hauser.

computer programs usually deal with data in chunks of multiple bits, and these chunks are commonly 8 bits in length, called *bytes*, the transmission rates and message lengths are often given in units of bytes per second and bytes, respectively. In that case, the transmission time formula is still the same. If one of the quantities is given using bits and the other given using bytes, a multiplication or division by 8 is needed. Other commonly encountered scale factors are milli-, micro-, and nano-, which are usually seen in the transmission time, and kilo-, mega-, and giga-used to scale the transmission rate or message length. Students should become familiar with the calculations of transmission time using all variations of these scaling factors.

The second contributor to the time taken to deliver a message is the time it takes for a signal to travel from the sender to the receiver. This is called the *propagation delay*. Calculation of propagation delay requires knowing the distance, d, between the sender and the receiver and the speed, v, at which signals travel between the two. Then, $T_{propagation} = d/v$. For electromagnetic signals in a free space, v is, of course, just the speed of light, 3×10^8 m/s. Signals in wires and optical fibers travel slightly slower, usually between half and three quarters of the free space speed. Again, scaling factors are commonly encountered in both the distance and the propagation speed. For terrestrial situations, rates in kilometers per second and distances in kilometers are typical, and the formula needs no adjustment. Students must be careful to distinguish transmission delay and propagation delay and apply each correctly as they may be easily confused.

Error rates may be expressed either as errors per bit, perhaps scaled as in errors per kilobit, errors per megabit, etc., or in errors per second. The two are inter-convertible provided the transmission rate is known. The error rates in modern terrestrial communication systems are quite low, but they are not zero, so system design has to provide a way to deal with errors.

Therefore, why do we care how long it takes to deliver a message? Students should reflect on this question in light of the earlier chapters. Consider how much data needs to be sent and the expectations that exist about what is to be done with the data. For example, if the data are being used to control a physical process, how often do the controls for the process need to be updated and how fresh must the data be?

We will see later that two additional factors contribute to message delay. First is the time required to process the data, at the sender, at the receiver, and at intermediate places a message may pass through. This is called the *processing delay*. Second, messages may sometimes have to wait for other messages to be processed or transmitted before they get their turn. Depending on how many messages are ahead of them, this *queuing delay* can represent a significant part of the total delay that a message encounters. Unlike transmission delay and propagation delay, queuing delay cannot be determined from basic properties such as transmission speed and distance in isolation. Rather, queuing delay is a property of the overall load on the system as well as its capacity. We will look more at queuing delay once we have discussed the other concepts in more detail.

4.2 Propagation Delay

Propagation delay is the time it takes a signal to travel over a distance. The time taken is, of course, governed by the distance that must be traveled and the propagation velocity of the medium carrying the signal. Computer communications are typically electromagnetic, and thus, the propagation speed is the speed of light, recall which is approximately 3×10^8 m/s under vacuum. (For our purposes, this is the most accurate as there are so many other factors that affect delay that we can at best estimate the delay anyway.) Communications can be carried out using other kinds of signals such as sound and even carrier pigeons(!), but that is rare, and if ever encountered that the analysis techniques are the same, just adjust the propagation velocity. Electromagnetic signals travel more slowly in physical media such as wires and optical fibers. The velocity is typically specified as the velocity factor (VF), which is a number between 0 and 1 that when multiplied by the speed of light under vacuum gives the speed of light in the medium. The VFs for network cables and optical fibers range from about 0.5 to 0.75. As a rule of thumb, if the exact specification is not known, 0.66 is both a decent approximation and a convenient one as it gives a propagation velocity of 2×10^8 m/s, which is easy to work with, even mentally. As can be seen from the above discussion, the line-of-sight propagation delay serves as a lower bound on the latency that can be achieved between two points. Of course, when dealing with real communications networks, wires and fibers seldom connect points over the line-of-sight path, so propagation delays will be larger. It is worth becoming familiar with order-of-magnitude estimates of propagation delays that may be encountered in real life. Over very short distances, you can estimate propagation delays as 1 foot per nanosecond, but estimates of propagation times for distances that may be encountered between substations and control centers and across distances associated with regions controlled by a utility, a balancing authority, or an entire grid are handy bits of knowledge for those involved in designing and using utility communication systems. (See exercises.)

4.3 Transmission Delay

Transmission delay is the second major component of overall communication latency. Transmission delay is associated with the rate at which information is transferred by the sender into the communication medium (and of course the rate at which it is transferred from the communication medium to the receiver). Transmission rates are given in bits per second, and so transmission delay for a message is determined by dividing the message's length in bits by the transmission rate in bits per second. As mentioned above, both message lengths and transmission rates are often specified with a scaling factor (kilo, mega, and giga), so adjust the calculation accordingly. Also, recall that in performing analyses related to transmission rates, it is essential to take into account which parts of the problem are specified using bits that are specified using bytes (8 bits).

Transmission rates encountered in various utility systems range from a few hundred bits per second for communication with analog modems over telephone lines all the way up to hundreds of gigabits per second using optical fibers. The transmission rates depend not only on the characteristics of the physical medium but also on the properties of the devices that drive the signals onto the wires or fibers. Thus, a particular kind of optical fiber might in some cases have a transmission rate of a few hundred megabits per second, while with different driving electronics and optics, it could support rates tens or hundreds of times larger. The choice of electronics and optics is mainly one of what will it cost to achieve the system goals.

Another way to look at this is to realize that the available transmission rate in a system constrains the amount of data that can be sent in a given amount of time. If system designs cut the margins for transmission rates too finely, it may prove difficult later to adapt to changing needs. Two examples come to mind from the past few years, the first concerning deployment of the so-called intelligent electronic devices (IEDs) in substations and the second concerning communication enhancements to mitigate concerns associated with cybersecurity.

Over the past two decades or so, a great deal of work has been done to equip substations with monitoring equipment such as fault data recorders and synchrophasor measurement units that monitor conditions on the grid at much higher rates than the Supervisory Control and Data Acquisition (SCADA) systems that were deployed over the prior decades. It has taken many years for the grid's communication infrastructure to begin to support the routine collection of the data produced by these devices and make them available for real-time monitoring and control of the grid. In this case, the large magnitude of the difference between data rates for SCADA and those for the newer IEDs typically requires a completely new data communication infrastructure. (See exercises.)

The situation with cybersecurity involved a less-dramatic difference in data volumes but rather a small increment that pushed many systems beyond their capabilities. The issue was that it became clear that protecting certain communications with encryption (for confidentiality) and digital signatures (to ensure that messages originated with the expected sender) was necessary for the security of the power grid. Unfortunately, many of the communications were carried over links that were already at their maximum capability, and the requirements for encryption and digital signatures added more bits that had to be transmitted. Therefore, transmission rate capabilities were an obstacle to making these important changes.

As we will see in the Section 4.4, having links with data transmission rates in excess of the expected load is also important in systems with multiple data sources in order to avoid excessive queuing delay.

4.4 Queuing Delay and Jitter

A given communication link can carry only a single message at a given instant. (While that is not strictly true for some technologies, it is a useful abstraction in that physical links that can carry multiple messages simultaneously can usually be regarded as a collection of single-message-at-a-time links.) If two or more messages are available to be sent and only one can be sent at a time, then at least one message must wait. The time spent waiting before transmission on a link begins is called queuing delay. Queuing delay can range from zero (in the case that there is no message already being sent when a new message is ready to send) to arbitrarily large amounts of time governed only by a device's available storage for holding messages waiting to be sent – that is, in the queue.

An analogous situation arises when you get in line (in the queue) to check out at a store. If the cashier is not currently serving someone else, you get served immediately. Your queuing time is zero in this case, although it still takes the cashier time to process your items, corresponding to the transmission time of a message. If there are others in the queue ahead of you, then you must wait for each of them to be served. The time from when you join the queue to the time you begin being served is your queuing delay.

Queuing arises when an outgoing link is fed messages from incoming sources whose aggregate transmission rates exceed the transmission rate of the outgoing link. In the case of a host computer attached to the link, the sources are programs running on the computer. In the case of a router or a switch, the sources are other links attached to the device. It should be clear that if incoming messages arrive at a rate higher than the transmission rate of the outgoing link, then the queue will grow ever longer. However, how does it work if the long-term arrival rate of messages is less than the rate of the outgoing link? Our experience with queues for cashiers at the store tells us that sometimes the queue will be longer and other times shorter, but our experience does not give us a very good handle on how to predict queuing delay.

Closed-form analysis of queuing delay is usually not possible for real-world traffic arrival patterns. A stochastic analysis for Poisson arrivals is, however, illuminating as to the behavior of queuing delays. Poisson arrivals are characterized by random, independent, identically distributed interarrival times that follow an exponential distribution. (You may wish to review your probability and statistics course.) Such arrivals are characterized by a single parameter, λ, representing the amount of arriving traffic per unit time – we could give it units of messages/second. The link has a service rate, μ, also measured in units of messages per second. Combining the two gives us the *utilization*, $\rho = \lambda/\mu$. As we observed earlier, if the utilization is greater than 1, the link is overloaded and the queue will grow arbitrarily long, only constrained by the available storage. We would expect that if the utilization is close to zero, then most messages will arrive to find the link not busy, so they will not wait in the queue and the average queuing delay will be brief. However, what about in between? With a bit of probability theory, it is not too hard to learn that the average overall time spent in the system (including the queuing delay and the service time) is $(1/\mu)(\rho/(1 - \rho))$. Focusing on that last term, we see that as the utilization approaches 1, the denominator goes to zero and the average time in the system goes to infinity.

This analysis of queuing delay even for this relatively simple and somewhat unrealistic configuration illuminates the earlier claim that having a transmission rate capability in excess of immediate need is not only a good defense against changing requirements but it also provides immediate benefit in the form of reducing queuing delays.

Recall that transmission and propagation delays are predictable from message lengths and link distances and that processing delays are typically much smaller than the other quantities, so we ignore them. Queuing delay, in addition to potentially adding a great deal of latency, also adds a component of variability to latency. That is, different messages may suffer different delays, and therefore, the time between messages' arrival at the receiver may vary. This variation is called *jitter*. Depending on the application, jitter may be a non-issue or it may be a rather serious concern. For example, consider watching a video or listening to an audio stream. If messages making up the stream do not arrive at the time they are needed for playback, the viewer/listener will experience a disruption. A programming technique called buffering may be employed to provide uninterrupted playback at the expense of increasing the delay seen by the viewer/listener, which is usually not a problem for recorded media, although it may be an issue for real-time viewing. Similarly, control applications need to be designed taking into account not only the average latency but also the maximum and minimum latencies that

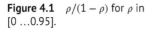

Figure 4.1 $\rho/(1-\rho)$ for ρ in [0 ...0.95].

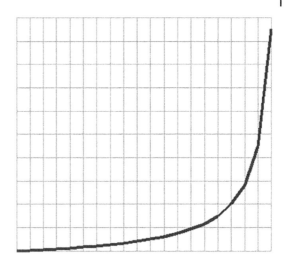

may be encountered in the system. In addition, this is difficult because of the potentially unpredictable nature of queuing delays (Figure 4.1).

4.5 Processing Delay

Until now, we have been discussing delays associated with communication links themselves. Communications have to traverse not only the communication links but also the equipment that connects the links – typically routers and switches. Routers and switches connect multiple links, and they decide for each incoming message which outgoing link or links the message should be sent on. The decision is based on the content of the message. The time to inspect the message and to look up the appropriate outgoing link(s) is called processing delay. Although processing delay is not zero, it can typically be measured in microseconds and thus is orders of magnitude smaller than propagation, transmission, and queuing delay for which milliseconds or even tens of milliseconds are typical. After ensuring that it is appropriate to do so, we are thus often, even usually, able to ignore processing delay in our analysis of overall communication delay.

4.6 Delay in Multi-hop Networks

As alluded to earlier, communication networks are typically made up of a large number of links connected by switches and routers. A message typically traverses many links, switches, and routers on its path from sender to receiver. Each traversed link is called a hop, and there is propagation, transmission, queuing, and processing delay associated with each hop.

There are several things to note about the performance of multi-hop networks. First, observe that the achievable end-to-end data transmission rate is limited by the transmission rate of the slowest link in the path. Further, observe that in real networks, the capacity of each link is shared among multiple senders, so each sender can only use a fraction of the available capacity. We will see in the next chapter how this sharing is controlled so as to avoid congestion collapse that occurs when senders attempt to send more data than the network is able to deliver.

The second thing to note about multi-hop networks is that different links can be busy carrying different messages at the same time. Thus, a multi-hop network does not behave identically to a single-hop network that covers the same distance, and the performance analysis needs to be adjusted to take this into account.

Finally, recalling that the variability of latency is mostly associated with queuing delay, one might expect that multi-hop networks, with multiple queuing delays, might offer some benefit in reducing jitter from the fact that each message encounters multiple random delays and that these might somehow average out resulting in less jitter. Unfortunately, I have never seen much benefit from this phenomenon and I suggest its investigation as a topic for exploration and research.

4.7 Data Loss and Corruption

Communication systems are not 100% reliable – there are multiple reasons why messages may be lost or corrupted. For example, electrical noise can interfere with data communication signals on both wired and wireless channels causing receivers to interpret the signals as containing different bits than the sender intended. This is, of course, potentially disastrous if decisions are made using incorrect information. Fortunately, with modern *error detection codes*, it is possible to ensure that corrupted messages are rejected by receivers, thus converting them to lost messages. (On exceptionally noisy channels, engineers may deploy *error correcting codes* to reduce the rate of message losses, but on many kinds of channels, this is not seen to be worthwhile as it comes at the cost of increased implementation complexity and it reduces the capacity of the channel.)

Corrupted messages are not the only source of message loss. Recall that in the earlier discussion of queuing delay, we observed that queue lengths are only limited by the amount of storage allocated to hold the queues. When the available storage is full, messages must be dropped as there is no place to put them. The design of the Internet, arguably the most important set of networking technologies developed to date, explicitly allows network devices to drop any message for any reason (or no reason). As we will see in the next chapter, other aspects of the Internet's design provide ways for receivers to detect dropped messages and signal to senders that they should be resent, if that is required.

It is worth noting that depending on the purpose for sending data, some data loss may be acceptable. For example, in an audio stream, occasional lost messages at the 1–5% level will not impede the overall intelligibility. Similar arguments can be made about some monitoring and control applications for the power grid: occasional lost messages are better dealt with by designing the applications to be tolerant of missing data. For such applications, data loss impacts the quality of the service delivered by the application, whether it is audio and video fidelity or system controllability.

On the other hand, many applications require that all of the data be delivered. For example, if you are looking at a web page, you would not be happy if some of text or parts of pictures were missing. For these applications, network systems provide ways to ensure that all of the data are eventually delivered, although this has costs, both in terms of implementation complexity and the amount of time that may be consumed recovering from data loss. In this kind of application, data loss essentially increases the communication latency and reduces the achieved data rates.

4.8 Summary

In designing and deploying communication systems to support power grid operations, there are many technology choices available. Each choice has implications for the overall performance of the system. In this chapter, we have covered the basic concepts: the components of overall delay, propagation, transmission, processing, and queuing, and the reasons for and implications of data loss. These concepts are essential knowledge for engineers, both as part of characterizing the performance needs of a given application and in designing networks that are able to support large numbers of applications with diverse needs.

4.9 Exercises

1   The ping command on Windows, Macs, and Linux sends a sequence of short messages to a destination computer. The computer echoes the message back to the sender, which then reports the time taken for the *round trip*.
Try the following commands:

```
ping localhost
ping www.google.com
ping <some~host~at~your~own~institution>
```

at a command prompt on your computer. (On Windows, the command will halt after sending a few messages. On Macs and Linux, type Ctrl-C to halt the command.)

localhost is the computer on which the ping command is running. We expect that a computer at your institution will be physically closer than www.google.com. Given this information, how do you interpret the times that you observe reported by the different ping commands? Do you think that the differences are mainly due to the differences in propagation delay, queueing delay, transmission delay, and processing delay? (It is entirely appropriate to consider these in combination and not in isolation!)

2   The traceroute command on Macs and Linux and the tracert command on Windows not only report the time taken to reach a destination host but also report the time taken to reach each of the intermediate routers along the path to that host. Note that each reported time in the output of this command corresponds to a separate message sent from your computer. Thus, it is entirely normal to see that the times for more distant sites along the path are sometimes smaller than what you see for closer sites. The usual cause is that different messages encounter different amounts of queuing delay, but it can even happen that different messages take different paths through the network. Another cause of seemingly contradictory output from this command is that intermediate routers may give, responding to this kind of traffic, very low priority, so processing delay may be significant. Also, note that some routers may refuse to respond to this kind of traffic in which case you will see a * in the output. In addition, some routers may refuse to forward this kind of Traffic, in which case, the output becomes just an unending sequence of *. Another feature of traceroute/tracert is that they report the names of devices along the path to the destination if it can figure them out. Sometimes, you can decode these to discover the city names and provider names along the path to the destination.

At a command prompt, use the traceroute (or tracert) command for each of the three hosts you used in problem 1. What do you observe? Choose several interesting features of each output, such as those mentioned above, and explain their significance.

3   A few years ago, Jim Gettys observed that as memory prices dropped, manufacturers of networking equipment found it easy and cheap to provide far larger amounts of space for queues than was ever considered when the Internet was being designed. What are the advantages and disadvantages of having large

amounts of space available for queuing? Research the topics of *Dark Buffers* and *Buffer Bloat* to get a sense of what impact these large buffers can have on the Internet as a whole. In the next chapter, we will address *congestion control*, which will lead to further understanding of the issue, but even with only the background of this chapter, you should be able to understand some of the issues.

4 SCADA systems that carry data from substations to control centers typically report the data at the rate of once every 2 seconds to once every 4 seconds. In contrast, modern substation intelligent electronic devices may collect and report data at rates up to several hundred times per second. Consider a substation that currently operates as part of a SCADA system that reports every 4 seconds over a 10000 bit/s link. If we upgrade the substation with IEDs that report the same data at 30 times per second, what link capacity will be needed?

5 Estimate the minimum propagation delay that messages will encounter traveling across a university campus (about 2 km), across a regional utility (about 200 km), and across one of the U.S. electrical grids (about 2000 km). Use a propagation speed of 2×10^8 m/s. How do these times compare to the ping times you observed in Problem 1. Remember that ping times are round trip, whereas your calculation here are one way.

6 What is the transmission delay for a 1 kilobyte message transmitted on a 1 megabit/second network link? (Do not forget to do appropriate scaling conversions including the ones for bytes to bits.)

7 If the utilization, ρ, of the link in the previous problem is 95%, estimate the average queuing delay. (Note that this is only an estimate as it is unlikely that interarrival times and services times are exponentially distributed.) What is the longest distance for which the estimated queuing delay exceeds the propagation delay (again, assume a propagation speed of 2×10^8 m/s).

8 Consider a situation in which a sender and a receiver are separated by 1000 km and connected with a link with a transmission rate of 100 kbits/s. We want to send 8 megabytes from the sender to the receiver. When sending a large amount of data, it is typically broken up into smaller chunks. For this problem, assume that the chunks are 1 *kbyte* in size. If the VF of the link is 0.66, how long will it take to deliver an 8 megabyte file over this link?

9 Now, suppose that there are 10 links like those of the previous problem connecting a sender and a receiver that are 10,000 km apart. How long will it take to deliver an 8 megabyte file in this situation again, assuming that the file is divided into 1 kbyte chunks. Why does it matter that the file is divided into chunks in this situation and not transmitted as a single 8 megabyte message that has to be completely received at each of the intermediate nodes before transmission can begin on the next link?

4.10 Questions

(1) Explain the different types of delay that contribute to the overall latency in communication of data packets between a source and destination.
(2) Pick a power system application that is deployed in a wide area, such as state estimation or voltage control. Discuss the latency requirements for this application, and the ways in which the latency can be managed to meet the required performance.
(3) Discuss the relation between queueing delay and jitter. What is their effect on smart grid applications?
(4) Pick a power system application that is deployed in a wide area, such as state estimation or voltage control. Discuss the effect of data corruption and data loss, and the ways in which they can be mitigated to ensure that the application meets the required performance.
(5) Compare and discuss the factors contributing to latency for (i) a power system application, (ii) on-demand video streaming over the Internet (such as YouTube or streaming services), and (iii) a cloud computing application. What are the similarities and differences between the three categories?

Further Reading

Bakken, D.E., Bose, A., Hauser, C.H. et al. (2011). Smart generation and transmission with coherent, real-time data. *Proceedings of the IEEE* 99 (6): 928–951. https://doi .org/10.1109/JPROC.2011.2116110.

Bush, S.F., Goel, S., and Simard, G. (2013). IEEE vision for smart grid communications: 2030 and beyond roadmap. *IEEE Vision for Smart Grid Communications: 2030 and Beyond Roadmap* 1–19. https://doi.org/10.1109/ IEEESTD.2013.6690098.

Hauser, C.H., Bakken, D.E., and Bose, A. (2005). A failure to communicate: next generation communication requirements, technologies, and architecture for the electric power grid. *IEEE Power and Energy Magazine* 3 (2): 47–55. https://doi.org/ 10.1109/MPAE.2005.1405870.

Kurose, J.F. and Ross, K.W. (2007). *Computer Networking: A Top-Down Approach Edition*. Addison-Wesley.

Yan, Y., Qian, Y., Sharif, H., and Tipper, D. (2013). A survey on smart grid communication infrastructures: motivations, requirements and challenges. *IEEE Communication Surveys and Tutorials* 15 (1): 5–20. https://doi.org/10.1109/SURV .2012.021312.00034.

5

Layered Communication Model

5.1 Introduction

In Chapter 4, we studied the performance of the communication infrastructure in the smart grid and the factors that affect it. In this chapter, we will study the models and frameworks that support the communication infrastructure.

Before the existence of the Internet, communication between computer terminals at remote locations was made possible by the ARPANET. ARPANET was the Advanced Research Projects Agency (ARPA) Network funded by the ARPA, an office in the United States Department of Defense. ARPANET was run on a framework of communication protocols called the Network Control Program (NCP). The NCP eventually gave way to the Internet's protocol suite, often somewhat incorrectly called TCP/IP, as the problem of scalability required a modular solution with well-specified boundaries between protocols in the suite. Protocols from the TCP/IP suite now underlie the communication capabilities of the Internet.

In the early 1980s the open systems interconnection reference model, OSI reference model, or simply the OSI model was created to keep up with the rate of technological change taking into account the need for higher levels of abstraction and the need for a standard that could accommodate future developments and competing interests. It was published in 1984 by the ISO and the International Telegraph and Telephone Consultative Committee (CCITT), as standard ISO 7498. OSI has two major components, an abstract model of networking, called the basic reference model, which is colloquially referred to as the seven-layer model, and a set of specific protocols. The OSI reference model was a major advance in the standardization of network operation as it promoted the idea of a clear separation in protocol layers and defines the interoperability between network devices and software.

Cyber Infrastructure for the Smart Electric Grid, First Edition.
Anurag K. Srivastava, Venkatesh Venkataramanan, and Carl Hauser.
© 2023 John Wiley & Sons Ltd. Published 2023 by John Wiley & Sons Ltd.

Although the OSI model and the TCP/IP model share similarities, the TCP/IP model eventually proved to be more popular and has more wide adoption. However, with the advent of cloud computing, and more upcoming sophisticated network functions that, it is believed that the OSI model might make a comeback. A comparison between the OSI model and the TCP/IP model is made at the end of the chapter.

In the smart grid with the advent of intelligent edge devices and the development of wide-area monitoring and control (WAMPAC), the Internet is becoming the future for power grid communication. One of the things that we always worry about in networking and data communication is the matter of performance. The major metrics for measuring the performance are delay, throughput, and loss, and we will be looking at the factors in networking that contribute to these aspects of performance. Delay, also referred to as latency, is the time for a message to traverse the network from the time that it is sent until it arrives at its destination. Throughput is how many messages per second are being delivered, which is measured in terms of bytes or bits per second, rather than messages per second. Loss is defined as the fraction of the data that will actually be arriving versus how much may be missing. We will also be covering the technologies and the trade-offs that exist between the technologies in choosing to deploy communications for power grid technologies such as the phasor measurement unit (PMU), which plays a critical role in WAMPAC applications.

PMUs are located in substations typically but may also be located in generation stations. The streams from the different PMUs are processed through often a phasor data concentrator (PDC), which then feeds upward into other PDCs. These may be located in control centers and even passed along with other entities in the power grid, such as transmission operators, reliability coordinators, and so forth. Along the way, the data may be used for operational purposes or may be stored for historical analysis. Hence, PMU data are typically used for many different purposes and that there are many PMUs that all relate to the same application of the power grid.

Before discussing the components of the communication frameworks, a few basic definitions need to be established.

1. *Hosts*: These are the end systems that are connected to the network and run network applications. They are also referred to as "end systems."
2. *Communication links*: These are the physical interconnections that enable hosts to communicate with each other. The common examples are fiber, copper, radio, and satellite links.
3. *Routers and switches*: Routers and switches are communication equipment that are used to move packets from source to destination based on certain rules.
4. *Nodes*: If we do not care whether a device is a router, a switch, or a host, we often just call it a node.

5. *Client server model*: In this model of communication, a client makes requests to an always available server. The common examples include web browsers and email servers that run on dedicated hosts.
6. *Peer-to-peer model*: This model uses minimal servers and instead relies on direct host-to-host communication.
7. *Protocol*: Protocols define the format and order of messages sent and received among network entities and actions taken on the messages such as transmission and receipt.

The roadmap for this chapter is to look immediately at the communication requirements for the PMUs and then dive into what is the Internet and what are the communication technologies that are used in the Internet? How do we talk about these things? How do we impose an intellectual framework around the Internet so that we could talk about pieces of the system in a coherent way but not have to deal with everything at once.

5.1.1 OSI and TCP/IP Models

In order to understand how networks fit together, it has become common to understand the networks as a set of layered protocols. In the early 1980s, when people were starting to do networking in a big way, they asked if there was any hope of organizing this structure in such a way that people could understand it piece by piece, instead of trying to understand it all at once and what they came up with was the idea of layered protocols. In recent years, there have been a number of discussions about whether layering is actually harmful to particularly the performance of networks, as it limits the functionality by requiring things to be rigidly layered. However, networking software is built around these protocol layers, and we allow exceptions to the layering in the implementation sometimes, but we definitely consider them exceptions rather than the norm.

The OSI and TCP/IP models are essentially a framework around a group of protocols that dictate how networks are set up. Both models are separated into various layers. It also creates modularization that eases maintenance and upgrades/updates on system components. Any changes in a particular module can be tested independently from the rest of the system, and its effect on the overall system becomes apparent easy. However, too much layering can also lead to excessive overhead that is sometimes considered harmful. This issue will be examined further in the comparison between the OSI and TCP/IP models.

The OSI model is a seven-layer framework with the following layers – (i) physical, (ii) link, (iii) network, (iv) transport, (v) session, (vi) presentation, and (vii) application. The OSI model is shown in Figure 5.1. The TCP/IP model on the other hand is simplified and combines functionalities from certain layers.

Figure 5.1 OSI layered network model.

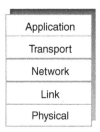

Figure 5.2 TCP/IP network model.

The TCP/IP model has the following layers – (i) link, (ii) network, (iii) transport, and (iv) application. The TCP/IP model is shown in Figure 5.2.

5.2 Physical Layer

The physical layer essentially deals with electrons and photons. It has to do with voltage levels, light frequencies, and other things related to the physical medium. From a software perspective, these considerations are abstracted away by the way that the physical hardware is built.

The physical layer deals with the transmission and reception of data between a device and a physical transmission medium. Once a message is received from the higher layers of the protocol stack, it converts the digital information into electrical, radio, or optical signals. Layer specifications define characteristics such as voltage levels, the timing of voltage changes, physical data rates, maximum transmission distances, modulation scheme, channel access method, and physical connectors. The protocol specification includes characteristics such as layout of pins, voltages, line impedance, cable specifications, signal timing, and frequency. Physical layer specifications are included in the specifications for almost all typical protocols such as Ethernet, USB, Bluetooth, and even industrial control protocols such as CAN bus.

5.3 Link Layer: Service Models

Hosts and routers are called nodes. The communication channels that connect nodes on a communication path are called links, which may be wired or wireless and sometimes will be local area networks, such as Ethernet and others. Other times, they will be point-to-point links.

A layer 2 packet is called a "frame" that encapsulates a datagram, which is received from the upper layer, which is layer 3. The data link layer has the responsibility to move a datagram from one node to an adjacent node over a single layer. It is important to realize that the datagrams are transferred across the network from one host to another by many different link protocols over many different links.

Therefore, the link layer services at a minimum constitute the following services:

1. Encapsulate the data received from the upper layer by adding a link layer header and a trailer.
2. Channel access, which is the process by which a shared medium is accessed by a particular host at a given time.
3. Link layer addressing, also referred to as MAC addressing.

MAC stands for media access control, and it is a misnomer because it does not relate to media access control, which is channel access. MAC addressing is just another kind of addresses that is used. It is important to distinguish the MAC addresses from the IP addresses, in the network layer, because they are of a different form.

The other link layer services that may or may not be present include (i) flow control, which is pacing between adjacent sending and receiving nodes so that it does not overflow buffers, (ii) error detection, which is understanding whether the electrical signal has deteriorated to the point it is not correct anymore, (iii) error correction, which will allow bit errors to be fixed by using additional redundant bits without relying on retransmission, and (iv) specification for half duplex or full duplex. Half duplex means the nodes at either end can transmit but not at the same time. Full duplex means the nodes at both ends for the link could be transmitting and receiving at the same time.

The link layer is typically implemented in a network adaptor or a network interface card (NIC). The NIC is attached to the host system and is a combination of hardware, software, and firmware.

5.3.1 Ethernet

Ethernet has become one of the most popular link layer protocols, and effective, modern implementation of the Ethernet protocols involves link layer switches. Wi-Fi protocols are also strongly influenced by Ethernet, so understanding Ethernet is essential for understanding the link layer in general.

Ethernet is connectionless; that is, there is no prior arrangement between the sending and receiving NICs. Sometimes, this connectionless nature makes transmission and reception on Ethernet unreliable. There are no steps taken to ensure that a message that is sent from one NIC is actually received at the other one, i.e. there is no positive or negative acknowledgment. On the other hand, the probability of loss is kept very low. To overcome this limitation of being unreliable at the network layer, we implement reliability in the TCP protocol. There are many different Ethernet standards for different physical layer media and different implementations, but they share some commonalities, such as the same form frame format, as shown in Figure 5.3.

In the various implementations of the Ethernet protocol, the central bus that is used to transfer messages has evolved from bus, to hubs, to switches. Routers on the other hand are devices at the network layer and will be explained later in the chapter. The bus is a communication medium to which all the hosts on the Ethernet are connected. A hub is a replacement of the bus and is also attached to all the hosts, and every bit that was sent by any of the hosts is received by all. An Ethernet switch is a smarter device than a hub. It implements its own transmitter and receiver for every host that is attached. It does store and forward on Ethernet frames, and it selectively forward frames to the outgoing links where they are needed.

A particular interest for the smart grid is building institutional networks with Ethernet switches, as shown in Figure 5.4. A large number of hosts can be connected onto an institutional network, up to thousands. Switches can be cascaded together to increase the number of devices if necessary, and switching

8 bytes	6 bytes	6 bytes	2	46–1500 bytes	4 bytes
Preamble/SFD	Destination address	Source address	Type	User data	FCS

Figure 5.3 Ethernet frame format.

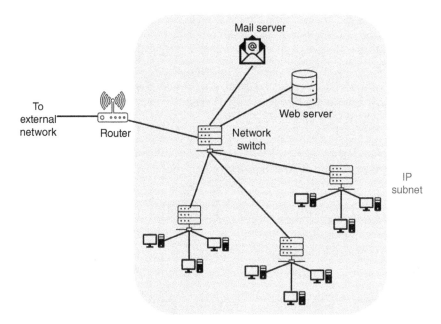

Figure 5.4 Institutional network using Ethernet.

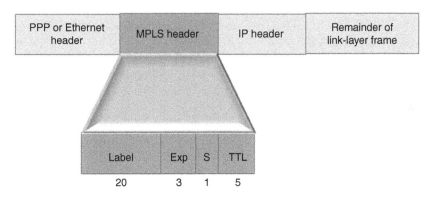

Figure 5.5 MPLS format.

tables continue to work as long as we do not introduce a cycle into this graph (Figure 5.5).

5.3.1.1 Link Virtualization

In the computing field, virtualization is common, with applications such as virtual memory, virtual devices, and virtual machines. An example of a virtual machine will be VMware or Docker, which implements a virtual machine within a computer to run other applications. The idea of virtualization is to abstract away the details of the lower layer and only deal with their promised levels of service. The idea behind link virtualization is to use technology from the higher layers such as the network layer to implement link layer functions, and the two most common ways this is achieved are through the asynchronous transfer mode (ATM) and the multiprotocol label switching (MPLS). ATM is a relatively expensive technology and primarily exists as an IP backbone that connects routers.

MPLS is another networking technology that has been incorporated in the Internet. It offers an advantage over Internet routers that it can direct traffic by just a single number that can be looked up in a table, as opposed to IP addresses that have to be searched for in tables. Therefore, MPLS potentially offers link layer switching much. It does require some prearrangement such as a signaling protocol to set up the forwarding in advance, but for engineering of traffic routes, MPLS is slick and works very well as part of an IP network.

5.4 Network Layer: Addressing and Routing

The network layer provides the functional and procedural methods for transfer of packets from one node to another connected to a network. Every node in the network is assigned an address, and the network permits nodes connected to it to transfer messages to other nodes by providing the data in a specific format and the address of the destination node. The networking layer is responsible for delivering the message to the destination node, possibly routing it through intermediate nodes. If the message is too large to be transmitted from one node to another on some link between those nodes, the network may implement message delivery by splitting the message into several fragments at one node, sending the fragments independently, and reassembling the fragments at another node. It may, but does not need to, report delivery errors.

In order to accomplish this mission of delivering data from anywhere to anywhere else, two closely related ideas need to be defined. The first is forwarding, which answers the question of which output link should a message be passed to when it arrives at a device. The second concept is routing, which determines how the forwarding process should be governed. Forwarding is this process of receiving a message, looking up the destination in the routing table or the forwarding table, and sending it onward to the next hop along the way to the destination. Routing is a set of algorithms that are used to build the forwarding table in each device. The routing algorithms are determined by the network service model that

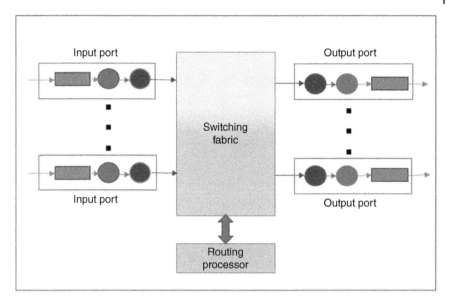

Figure 5.6 Overview of a router.

are designed and implemented. These service requirements can be enforced on individual datagrams or on a flow of datagrams. For an individual datagram, the service model can enforce requirements such as guaranteed delivery, or go further and enforce guaranteed delivery within a particular time bound. For a network flow, the requirements could be to ensure that the datagrams are delivered chronologically, or have a minimum bandwidth for a particular flow, or restrictions on interpacket spacing.

An overview of the router architecture is shown in Figure 5.6. A router has a number of input ports, and each port typically may or may not have a queue associated with it. The packets arrive in the queue, and they get passed through a switching fabric that moves the data from the input port to the output port, and the behavior of the switching fabric is controlled by the routing processor.

The IP protocol defines how messages are laid out, what the headers and trailers are, and what rules are to be followed when things malfunction? The format of an IP datagram involves a 20 byte header that is attached in the front of any data that is received from the upper layer, namely, the transport layer. The 20 byte header contains the following information as shown in Figure 5.7.

The IP datagram can be of a length up to 64 kB. However, link layer protocols such as Ethernet only support lengths up to 1500 bytes. To account for this mismatch, IP has an ability to fragment and reassemble large packets.

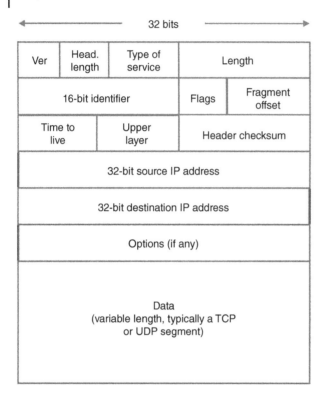

Figure 5.7 IP datagram format.

5.4.1 IP Addressing

An IP address is a 32-bit identifier that is tied to a particular network interface. It is also sometimes referred to as a host address, but in reality, a host may have more than one network interface and hence more than one IP address. The convention for writing the 32-bit addresses is that we break it up into 8 bits (1 byte), and the value of each of the bytes is specified in decimal with a "dot·" between. This method of referring to IP addresses in decimals is a convention used to make them easier to be read by the users. To achieve the movement of packets between IP addresses, an addressing structure is created, called the subnets. Several hosts can all exist on the same link layer network; those hosts altogether are called a subnet, and the subnet itself has a subnet address. The subnet address is the high-order bits of the 32-bit IP address. The low-order part of the 32-bit quantity is called the host part or a host address. In addition, the subnet part and the host part together give the complete IP address. Figure 5.8 shows a network containing three subnets with different hosts in the same subnet highlighted.

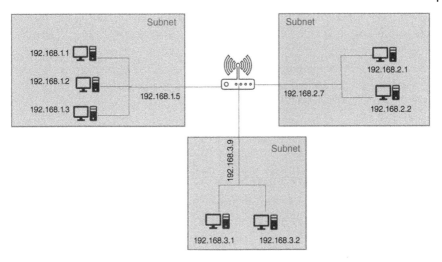

Figure 5.8 IP addressing – subnets.

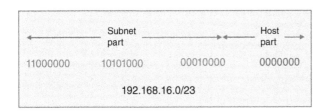

Figure 5.9 IP addressing – host part and subnet part.

This is an example of a complete IP address with the subnet part and the host part divided as shown in Figure 5.9. This subnet permits up to 512 hosts, of which 510 addresses are available for devices. The 11111111 (all 1's) host address is used for broadcasting on the subnet and the 000000000 (all 0's) part is used when a host does not yet know what its proper address is.

There are several methods by which a host gets assigned an IP address. The old-fashioned way is that a system administrator manually assigns an IP address to each device on the network. The other method is to dynamically allocate IP addresses, which is achieved through a protocol called the Dynamic Host Configuration Protocol (DHCP). When a host joins a network, it uses the link layer to broadcast a discover message. The discover message goes to all the hosts on the network. A host on the network is assigned to be a DHCP server, and it will respond with an offer message that contains an IP address that the host might be able to use. The host then requests assignment to that IP address, and the server sends back an acknowledgment. The addresses are assigned typically between a few hours and a few days, unless dictated otherwise by the system administrator. If a host wants to continue using the IP address, it has to go through a renewal protocol, which is also part of DHCP. The Internet Corporation for Assigned Names and Numbers (ICANN) is responsible for ensuring that the IP addresses being allocated are unique without any duplicates being assigned to internet service providers (ISPs).

5.4.2 Routing

The job of the routing algorithm is to fill in the forwarding table of each router so that the packets, when delivered to the router and forwarded according to the forwarding table, eventually reach their destination. To design these algorithms, the common practice has been to model the network as a graph consisting of a set of nodes for routers or the subnets, and edges being the links that connect the subnets. Usually, routing algorithms abstract away the hosts connected to the subnet and only worries about the routing between subnets.

In Figure 5.10, the routers are the lettered circles and the links are the numbered lines. Routing algorithms try to solve the question: What is the least cost path between a source and a destination? The cost in this question can be designed by the user, such as minimizing the number of hops between the source and the destination. There are a couple of different classifications of routing algorithms depending on whether the route computation is done using full knowledge of the network.

1. Link state algorithms: Routers have complete topology information
2. Distance vector algorithms: Distributed algorithms where routers have knowledge of physically connected neighbors and their information and exchange information with neighbors.

5.4.3 Broadcast and Multicast

In the power grid with PMUs that are producing measurements at a rate of 30 times per second, it is very likely that there are going to be multiple users of any given data stream. The idea of broadcast is that a source sends a packet into the network and the packet is delivered to all the other nodes in the network. There are a couple of ways to achieve this as shown in Figure 5.11: (i) have the source itself duplicate the packet and address a separate copy to every one of the other nodes and (ii) have the network duplicate the packet.

Source duplication is wasteful of bandwidth on the link adjoining the source. The source also needs to know the addresses of all the destinations that may want to receive that data.

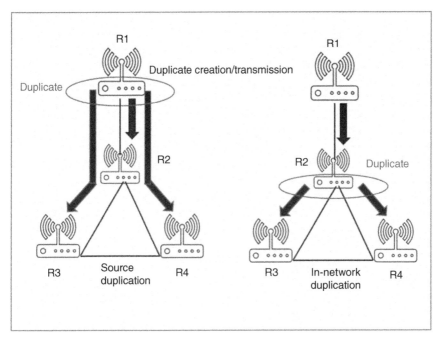

Figure 5.10 Routing algorithms set up.

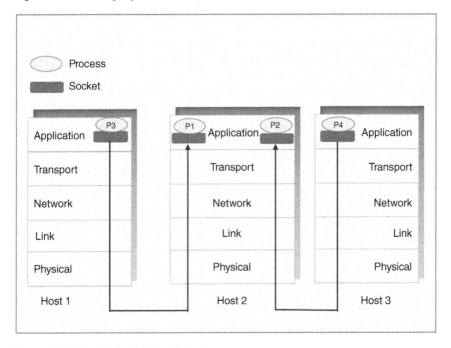

Figure 5.11 Packet duplication strategies.

In addition, both of these can be addressed by some scheme that allows the network to do the duplication of. Data are needed in order to reach destinations that are also present in the network. These issues can be solved by network duplication, but we also need to design network duplication carefully to avoid problems such as broadcast storm, where the same packet is being repeatedly forwarded across the network. This is addressed by creating spanning trees, where the nodes are connected without having any cycles in the graph. Multicast is a refinement on the idea of broadcast in which we have several nodes that require the data and a number that do not. The goal of multicast is to find a tree (or trees) that connects the routers that have the destination of the data in their local area networks or that need to forward the data from the source to one of the routers that has a destination for the data. Multicast and broadcast are very useful for smart grid applications, where the data generated by field devices need to be sent to multiple destinations including historians, control applications, and Human Machine Interfaces. IEC 61850, which is a smart grid protocol used for substation automation, has two multicast protocols: Generic Object-Oriented Substation Events (GOOSE) and Sampled Measured Value (SV). These protocols are used to collect information about the status of the network in real time, to update the status of intelligent electronic devices (IEDs), and to retransmit control commands.

5.5 Transport Layer: Datagram and Stream Protocols

The transport layer provides the functional and procedural means for transferring variable-length data sequences from a host program to a destination program, which may or may not be in different hosts. Transport layer protocols also provide specifications on the quality of service, which vary widely based on the protocol used. An important distinction to note is that the transport layer provides program-to-program communication, while the network layer provides host-to-host communication. The transport layer may control the reliability of a given link through various functions such as flow control, segmentation/desegmentation, and error control. Certain transport layer protocols are connection oriented, which means that the transport layer can keep track of the segments and retransmit those that fail delivery, and may also provide the acknowledgment of the successful data transmission and send the next data if no errors occurred. The transport layer creates segments out of the message received from the application layer if the message is longer than the datagram size.

The services offered by the transport layer have evolved with the evolution of the Internet. However, the services offered and their implementation may be the most ideal way to do things, especially when it comes to serving the needs of control systems for the smart grid. There might be other ways to do things that would be better, but the economics of the situation is that using Internet-based protocols is the dominant approach, and it would be too expensive to start from scratch. Other implementations may emerge in the future, or the Internet will adapt to the needs of such uses.

There are two principal variants of transport layer protocols: (i) transmission control protocol (TCP) or (ii) user datagram protocol (UDP). The principal function of TCP is to provide reliable, chronological (in-order) delivery of streams of bytes to applications. Secondarily, TCP implements congestion control and flow control and requires a connection setup phase in order to do this. The unreliable unordered delivery transport layer protocol is UDP, and it provides very little beyond basic IP (Figure 5.12).

An application sends data down through the stack and makes use of the physical network, and for an incoming packet, the transport layer has to figure out which application that data belong to. This process is referred to as multiplexing and demultiplexing. This is achieved through an operating system concept called "socket," which provides a logical endpoint for applications or processes to access physical resources. The transport layer protocol also creates a logical construct called "port," which is used to label various processes. The operating system is responsible for mapping sockets to ports and vice versa.

In order to implement this, both TCP and UDP include in their headers a 16-bit source port number and a 16-bit destination port number, as shown in Figure 5.13. The IP datagrams carry one transport layer segment, and each segment has one of

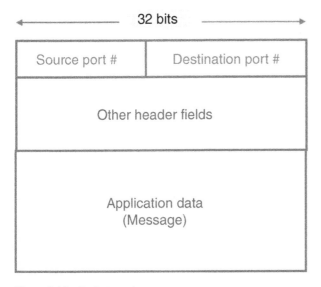

Figure 5.12 Sockets and processes.

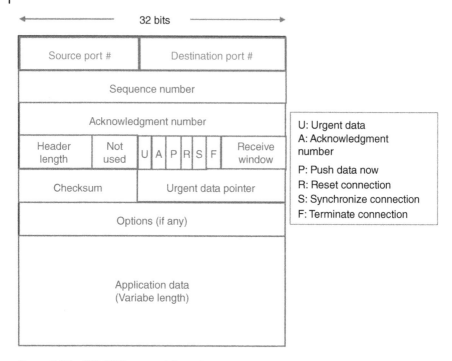

Figure 5.13 TCP/UDP segment format.

these source ports and destination ports. Rules are implemented in the transport layer and in the operating system to figure out which socket a particular datagram belongs to.

5.5.1 UDP

UDP can be considered a "best-effort IP," as the only additional information that it carries relates multiplexing and demultiplexing of datagrams. Segments may be lost, or delivered out of order, and there are no guarantees about timing or anything else. The advantages of UDP are that the sender and receiver do not need to have any communication before beginning sending data. In addition, every UDP segment is handled independently. A consequence of this is that UDP is better suited for multicast whereas TCP does not.

If an application requires reliable data transfer using UDP, it is up to the application to implement it. The UDP checksum is a very simple checksum, and it is not really capable of detecting deliberate manipulation of the data. Secure versions of UDP have been proposed based on more complex cryptographic mechanisms for various applications.

5.5.2 TCP

TCP is a point-to-point protocol. Its goal is reliability in delivery of byte streams, and there are no message boundaries in TCP. If an application needs to identify that one message stops and another one starts, it is up to the application protocol to define how that is achieved. TCP implements pipelining so that it can achieve good utilization of links that have high bandwidths and long delays. It requires send and receive buffering in the hosts in the end hosts but not in the routers as the routers are network layers only and do not participate in the transport layer. TCP implements full duplex data transfer and is able to send data in both directions on a single connection, and it has a maximum segment size that is identified for each connection. In order to implement the reliable transfer, TCP requires a connection setup phase at the beginning in which initial messages are exchanged in order to set up the sending and receiving state. TCP implements flow control, which means that the sender will not overwhelm the receiver. If the receiver's buffers overflow, the TCP protocol will cause the sender to slow down (Figure 5.14).

TCP has a header field that is added to the data coming down from the application, and it consists of 16-bit port numbers for the source and destination.

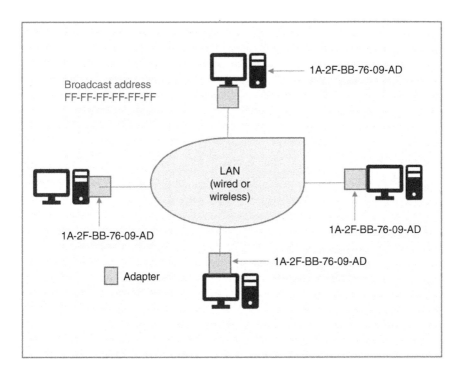

Figure 5.14 TCP segment structure.

The source and destination IP addresses meanwhile are part of the IP header. There is a 32-bit sequence number and a 32-bit acknowledgment number that count the number of bytes that have been sent on the connection. There are several flags that are associated with initiating connections, which are **RST, SYN, and FIN**.

TCP connections are initiated by the client, as a client is typically a process on a host that may require network connection while the server is usually static in its location and availability. Therefore, the client initiates a connection by asking for a new socket that is connected to a particular host name and port number. The server lets the networking system within its operating system know that it is willing to accept the connection by making a call on a system called "accept," and when those two things are done, eventually, the result is that across the network, the client sends a synchronize segment or a **SYN** segment to the server. The client picks a random initial sequence number, which is a way of avoiding certain kinds of attacks. The server responds with a **SYN ACK** segment that has both the SYN flag bit and the ACK flag bit set in it. It allocates its buffers and it specifies its initial sequence number. When the client receives the **SYN ACK**, it is allowed to start sending data. This connection initiation and setup process is called the three-way TCP handshake. Similarly, at the end of the connection, either the client or the server sends a **FIN** message that terminates the connection and frees up the resources. It is important to distinguish between flow control and congestion control, especially in the context of TCP. Flow control is a mechanism in TCP by which the sender's sending rate is restricted to ensure that the receiver is not overwhelmed. Congestion control is a mechanism to ensure network "fairness" in a scenario where multiple applications aim to utilize the same network resources, and congestion control is used to ensure that the resources are distributed in a fair manner between the competing applications. This is usually achieved by reducing the amount of sent packets to a queue, which is usually at the router. Flow control is used to manage resources at the receiver side, while congestion control (which is usually initiated by a network level device such as the router) is used to manage the network resources.

Comparing TCP with UDP, TCP offers a lot of advantage with reliable communication, but this does come at a cost. For a smart grid application, the user might be more interested in the data at the current time, rather than worrying about retransmission of data from a few seconds before. Those few seconds might be critical in control centers when immediate situation awareness becomes critical for scenarios such as power grid oscillation monitoring and control or other such transient behaviors. Hence, the choice of using TCP against UDP needs careful consideration.

5.6   Application Layer

An application layer is an abstraction layer that builds on top of the layers that we have studied so far and specifies the shared communications protocols and interface mechanisms to be used by hosts for a particular application. There are a variety of application layer protocols and applications, including everyday protocols such as HTTP, HTML (or email), and FTP, or a peer-to-peer application such as the BitTorrent protocol. Domain Name Service (DNS) is a protocol and is also an application layer protocol that is integral to the functioning of the Internet but can be considered as an example of a protocol that lies outside the strict layer structure.

The Internet has revolutionized the idea of distributed computing in that by operating above the transport layer and in turn above the network layer, we are able to program distributed applications in such a way that they do not have to do anything at all to the network core. The network core has many moving components, and if individual applications need to specify all operating parameters over the network core, that would be very inefficient. Recent research in the area of software-defined networks (SDN) is an exception in that it allows end user applications to make changes to the network core.

There are three primary application architectures:

1. Client server architecture
2. Peer-to-peer (P2P)
3. Hybrid

The client server architecture is one where there is an "always-on" host that is the server, and it typically has a permanent IP address. The services that run on the server typically have well-known port numbers associated with them. This makes it easy for the clients to get in touch with the server and make use of the services that they provide. In this architecture, the clients do not communicate directly with one another, rather they communicate through the server.

In a peer-to-peer architecture, the end systems communicate with each other without ever using servers. The end systems come and go on the network, and there is a mechanism for the end systems to find each other in which it would be a completely distributed implementation, but it is also very common to have a hybrid situation in which there is an always on server that helps the end system clients to find each other. Once they have found each other, then they communicate directly.

Each application layer protocol defines the types of messages that are going to be exchanged in that protocol. It defines exactly (i) what the fields in the messages mean, (ii) how they are separated from one another, (iii) what the messages mean, and (iv) the rules for when and how processes must send and respond to messages. Protocols are typically defined in standards, and these standards are either available publicly or are made available to user groups in advance to ensure standardization across various applications.

5.7 Glue Protocols: ARP and DNS

One of the problems that must be solved at the link layer is how to ensure that messages sent from one host get to the correct destination host. This is the topic of addressing, and when we are dealing with the link layer, it is called link layer addressing.

The link layer address is also known as the MAC address, LAN address, physical address, Ethernet address and is used to direct a frame from one interface to another on the same link layer. A network address on the other hand is used to get datagrams from one host on the Internet to the correct destination network. MAC addresses are typically 48 bits long and are unique worldwide. This is a feature provided by the IEEE, and a manufacturer of networking equipment buys a portion of the overall address space and assigns one address per device (Figure 5.15).

These 48-bit numbers are written as 6 hex bites, which turn into 12 hex digits. A hex digit is 4 bits, and so 12 hex digits represent 48 bits and use dashes or colons to separate the digits.

Now, the problem arises – given an IP address, how do we go from one of those to the MAC address? One advantage is that we only need to know the MAC address

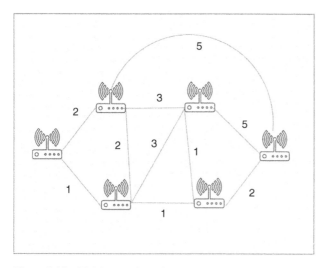

Figure 5.15 Link layer addressing.

at the link layer, as the IP address is sufficient to ensure that the message reaches the right network. In the link layer, if we want to send to a particular host on LAN, we have to know the appropriate MAC address. To discover the MAC addresses in a LAN, the address resolution protocol (ARP) is used. Every node in the LAN has a table in which it records the correspondence between IP addresses on the LAN and MAC addresses. It will not be for all the addresses on the LAN because it only cares about the ones that it needs to send to. Therefore, it will not record the ones that it does care about, and it typically remembers these associations only for a small period because MAC address assignments can change.

Suppose A wants to send a message to B and A does not know B's MAC address. A broadcasts a query to all the machines on the LAN saying what MAC address corresponds to B's IP address. All the machines on the LAN see that query, but only B knows the answer to the question. B receives the query packet and replies with its own MAC address. Therefore, by sending it to A's MAC address, in the process of doing this, B learns A's MAC address and A learns B's MAC address. ARP is completely plug and play, and it does not require any administrative support to make this work because of the way it is designed.

5.7.1 DNS

Domain name is the concept of assigning human sensible names (such as www .google.com) to a 32-bit IP address. The Domain Name System (DNS) is used to resolve the IP address of these hosts on the network to a machine-sensible 32-bit IP address. DNS is a distributed database that is implemented at different levels of hierarchy, and there are many different administrative domains (the various "dot" domains, such as.com or.edu) that are associated with maintaining it. These domains are usually maintained by a small number of organizations, each responsible for a specific domain.

DNS runs as an application layer protocol. It uses UDP and TCP in different parts of its operation and is able to answer queries from hosts using either TCP or UDP. It uses the transport layer protocols to exchange information among the DNS servers as well. DNS is an example of one of Internet design philosophies where the network core operation is kept simple, but the complexity is pushed out to the edge. At the network core, hosts interact based only on the host names, but DNS is a service that is deployed at the edge, which significantly enhances the usability of the network.

DNS offers various services including

1. Hostname to IP address translation
2. Host aliasing (offering aliases to hostnames)
3. Mail server aliasing

4. Load distribution (distribute load across multiple servers or IP addresses for one canonical name/domain)

DNS needs to be a distributed system and should not be a centralized service as it would have a single point of failure, and it would also have a tremendous problem scaling to serving billions of hosts in the network. It would also be very difficult to manage such a vast database. DNS queries are not performed every time a host needs to contact a server; typically, a fair amount of caching is done at the host level to store most commonly accessed domains.

5.8 Comparison Between OST and TCP/IP Models

Despite the TCP/IP model using a different methodology for layering than the OSI model, these layers are often complementary with the OSI model.

1. The application layer of the TCP/IP model maps to the OSI application layer, presentation layer, and the session layer.
2. The TCP/IP model's transport layer maps to the OSI transport layer, as well as incorporates transmission control functions such as the graceful close function of the OSI session layer.
3. The TCP/IP's link layer corresponds to the OSI data link layer and may also include functions as the physical layer, as well as some protocols of the OSI's network layer.

A difference between the TCP/IP model and the OSI model is in the treatment of routing protocols. The OSI routing protocol IS-IS is part of the network layer and does not depend on CLNS (Connectionless-mode Network Service) for delivering packets from one router to another but defines its own encapsulation. In contrast, the inter router communication in the TCP/IP model that uses OSPF, RIP, BGP, and other routing protocols are transported over IP, and the routers act as hosts for this purpose.

Although the OSI model is often still referenced, the TCP/IP model has become the standard for networking. TCP/IP offers an easier approach to computer networking, making independent implementations of protocols simpler.

5.9 Summary

In this chapter, we studied the basis of communication networks, with detailed discussions on the various layers of the TCP/IP model, which is the backbone for networking as used today. Commentary on how these layered protocols are used in the smart grid is also provided, with examples. Finally, the difference between the TCP/IP model and the OSI model is briefly discussed.

5.10 Problems

1 What does MAC stand for in the link layer?
  A Media access channel
  B Multiaccess channel
  C Media-authenticated communication
  D Media access control
  E Multiaccess control

2 What does it mean to be a "connectionless protocol," such as Ethernet?
  A There is no prior arrangement between the sender and receiver
  B It is used to send one message at a time
  C It can be used regardless of whatever communication medium is used
  D Devices can communicate with any addressing requirements
  E They do not need any other network devices such as switches

3 How is this common IP address 01111111.00000000.00000000.00000001 more typically represented?
  A 192.168.0.1
  B 192.0.0.1
  C 124.0.0.1
  D 10.0.0.1
  E 127.0.0.1

4 For a network address of 192.168.1.0 with a subnet mask 255.255.255.0, what are the total number of hosts in the network, and how many can be assigned by the system administrator (i.e. usable hosts)?
  A 255, 255
  B 256, 254
  C 254, 253
  D 254, 252
  E 256, 256

5 Web browsers (such as Google Chrome, Mozilla Firefox, or Microsoft Edge) typically use what type of application layer architecture?
  A Peer-to-peer
  B Hybrid

C Client server
D Domain Name System (DNS)
E Border Gateway Protocol (BGP)

5.11 Questions

(1) Describe the layers in the OSI network model and the TCP/IP network model. How are they different?
(2) Briefly explain how institutional networks can be created using link layer technologies.
(3) How does routing work? Where are routing algorithms typically placed in the TCP/IP network model? What are some typical routing algorithms?
(4) Discuss broadcast and multicast, specifically looking at their applications for the smart grid.
(5) Pick a power system application from the discussions in the previous chapters and discuss which would you use, TCP or UDP, for the application that you choose? Explain the advantages and disadvantages.

Further Reading

Bush, S.F. (2014). *Smart Grid: Communication-Enabled Intelligence for the Electric Power Grid*. Wiley.

Galli, S., Scaglione, A., and Wang, Z. (2011). For the grid and through the grid: the role of power line communications in the smart grid. *Proceedings of the IEEE* 99 (6): 998–1027.

Mahmood, A., Javaid, N., and Razzaq, S. (2015). A review of wireless communications for smart grid. *Renewable and Sustainable Energy Reviews* 41: 248–260.

Meyer, D. and Zobrist, G. (1990). TCP/IP versus OSI. *IEEE Potentials* 9 (1): 16–19. https://doi.org/10.1109/45.46812.

Panek, C. (2020). Defining networks with the OSI model. In: *Networking Fundamentals*, 43–73. Wiley. https://doi.org/10.1002/9781119650768.ch2.

Saadawi, T.N., Ammar, M.H., and El Hakeem, A. (1994). *Fundamentals of Telecommunication Networks*. Wiley-Interscience.

6

Power System Application Layer Protocols

In this chapter, we will aim to understand the uses and implementations of data communications in the smart grid, specifically the power system protocols that support various applications. A wide variety of requirements exist for various applications, and several approaches have been proposed to meet these requirements. In this chapter, which is an extension of Chapter 5, we will build on the TCP/IP stack and look more closely at the application layer protocols used in the smart grid. Specifically, we will address Supervisory Control and Data Acquisition (SCADA) and Distributed Network Protocol 3 (DNP3) protocols, communication between control centers using ICCP, the phasor measurement unit (PMU) protocol C37.118, substation communications in IEC 61850, smart meter communications, and the time synchronization methods for these applications.

6.1 Introduction

SCADA systems are essential for industrial control systems such as the smart grid. They are also ubiquitous in other domains such as large factories and manufacturing facilities. There are two components to SCADA system – data acquisition and control. Data acquisition, at least in the smart grid area, is usually slow and is in the range of 2–4 seconds. There are also mechanisms where data acquisition happens based on a pre-defined trigger such as a critical event. The data are transmitted from the field sensors using a communication layer and are recorded in the Historian Database, which is usually located in a central location such as the control center or the substation. The data from the SCADA system are processed for use by various power system applications that form the energy management system (EMS), and the system operator interacts with the data through a Human–Machine Interface (HMI).

Cyber Infrastructure for the Smart Electric Grid, First Edition.
Anurag K. Srivastava, Venkatesh Venkataramanan, and Carl Hauser.
© 2023 John Wiley & Sons Ltd. Published 2023 by John Wiley & Sons Ltd.

6.2 SCADA Protocols

A typical system has the following components as shown in Figure 6.1:

1. **Operator** is usually a human driving the control of the system and is responsible for monitoring the performance of the system, defining tolerances for triggering alerts, addressing alters, and taking control actions as necessary. The operator's function can also be augmented with smart or automatic control. The operator can either be located physically on-site or perform their function from a remote location through suitable access controls.
2. **Human–Machine Interface (HMI)** is the interface that allows the operator to interact with the SCADA system. The HMI is usually a software interface that

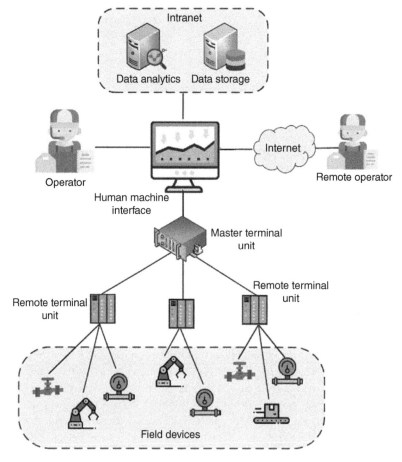

Figure 6.1 A typical SCADA system architecture. Source: Adapted from Pliatsios et al. [2020].

translates the data coming in from the sensors to actionable information and translates the operator commands to actuation signals to the control devices.

3. **Master terminal unit** (**MTU**) is the component that gathers the information from the sensors or remote terminals deployed in the system and transmits them to the HMI. It also provides the high-level control logic for the system.

4. **Remote terminal unit** (**RTU**) is the component that interacts with the MTU to send measurements to the HMI and is responsible for communicating the control signals to the field devices.

5. **Field devices** include devices that are responsible for sensing and measuring the quantities of interest and devices that generate actuation signals for field equipment. These included devices such as meters, temperature sensors, or actuation equipment such as controllers.

The SCADA systems utilize the same TCP/IP network stack but have a variety of application layer protocols to account for various types of devices, processing capabilities of different components, and communication and computation requirements of the application that it supports. The application layer SCADA protocols can be broadly grouped under four categories:

1. Fieldbus-based protocols such as BITBUS and PROFIBUS
2. Ethernet-based protocols such as DNP3 and IEC 61850
3. Serial-based protocols such as IEC 60870 and Modbus
4. Common Industrial Protocols (CIP) such as DeviceNet, ControlNet, and Highway Addressable Remote Transducer (HART)

A list of these application layer protocols, the network infrastructure on which they are run, typical communication topologies used, data rates, and their maximum distance is presented in Table 6.1 (Pliatsios et al. [2020]).

Of these protocols, the most common ones used in the smart grid area are DNP3 and IEC 61850. Other application layer protocols are also used in the smart grid, such as IEEE C37.118 that is the PMU protocol; however, they have different data rates and do not fit under the SCADA classification.

6.2.1 DNP3 Protocol

The DNP3 is a SCADA protocol that enables communication between components in a process automation system and facilitates data exchange between monitoring and control systems. In SCADA systems, it is used for communication between MTU, RTUs, and field devices, which are also referred to as intelligent electronic devices (IEDs). The DNP3 protocol is also called the IEEE 1815 standard and was originally proposed in 1990. It follows a layered architecture as shown in Figure 6.2.

Table 6.1 Application layer SCADA protocols.

Protocol	Network infrastructure	Topologies	Data rates	Maximum distance
BITBUS	Fieldbus	Bus	62.5 Kbps, 375 Kbps, 1.5 Mbps	1200 m
DC-BUS	2-wire cable	Line	115.2 Kbps up to 1.3 Mbps	100 km
Distributed network protocol 3	Ethernet	Line, peer-to-peer	100 Mbps, 1 Gbps	100 m
EtherCAT	Ethernet	Ring, line, daisy-chain	100 Mbps	100 m
Ethernet powerlink	Ethernet	Tree, line, star, peer-to-peer	100 Mbps	100 m
Foundation fieldbus H1	Fieldbus	Point-to-point, bus with spur, daisy-chain, tree	31.25 Kbps	1900 m without repeater, 9500 m with up to 4 repeaters
Foundation HSE	Ethernet	Tree, line, star, peer-to-peer	100 Mbps	100 m
HART	2-wire cable	Point-to-point, multi-drop	1.2 Kbps	3 km
IEC 60870	Serial, Ethernet	Ring, tree, line, star	N/A	N/A
IEC 61850	Ethernet	Ring, tree, line, star	N/A	100 m
Modbus	Serial, Ethernet	Line, star, ring, mesh (MB+)	100 Mbps, 1 Gbps	N/A
PROFIBUS	Fieldbus	Point-to-point, bus with spur, daisy-chains, tree	9.6 Kbps to 12 Mbps	100–1200 m, 15 km for optical channel
PROFINET	Ethernet	Ring, tree, line, star	100 Mbps, 1 Gbps	100 m
RAPIEnet	Ethernet	Line, ring	100 Mbps	100 m
SERCOS III	Ethernet	Line, ring	100 Mbps, 1 Gbps	N/A
Unitronics PCOM	Serial, Ehternet	Ring, line, star	100 Mbps	100 m
WorldFIP	Fieldbus	Bus	31.25 Kbps, 1 Mbps, 2.5 Mbps, 5 Mbps	1 km

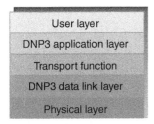

Figure 6.2 DNP3 network stack.

The lower level is a physical layer that interfaces with the physical media, such as copper, radio, etc. The data link layer has 256 byte (max) frames, with checksums, addressing, and acknowledgments. Moreover, it handles additional functions such as data link frame synchronization, flow control, error handling, and link status probing. The transport function is considered a sublayer of the application layer and fits above the data link layer. It is also considered to be a pseudo-transport layer and is responsible for breaking the DNP3 fragments into segments. A transport layer segment is composed of the header and the application data. The header is composed of the FIN, FIR, and sequence fields, which indicate whether it is the final or first fragment, and a variable, which is used to differentiate subsequent fragments. The application layer provides the master–slave functionality and also enables the use to have specific application level acknowledgments. The user layer contains a user code that can be tweaked by the user for specific applications.

The DNP3 protocol can operate over a variety of configurations such as slow serial links or multi-drop links (the same information sent to multiple slaves), concentrated links (communicated through a data concentrator), and hierarchical links. It should be noted that the DNP3 protocol has several deficiencies in the context of smart grid applications:

1. The DNP3 protocol does not support routing in its default implementation.
2. It does not have any security features, although improvements have been proposed in secure DNP3.
3. It is designed for slow links and for polling from field devices.
4. It has inflexible data naming and addressing conventions, and protocol objects must be manually mapped to power system functions for each different device, application, and vendor.
5. It does not support peer-to-peer communication and focuses mainly on delivering information to the control center.
6. Measurements are time-stamped at the time of receiving and not when the measurements are made.

6.2.2 IEC 61850

IEC 61850 is a suite of protocols for substation automation that is fast becoming the standard for smart grid applications across North America, Europe, and Asia. It was proposed and designed to accommodate future smart grid applications and hence tries to cover all needs for substation automation. It also specifies a database and naming convention, thus addressing problems with legacy protocols such as DNP3. The 61850 protocol can be used across the layers, enabling communication from the SCADA HMI to the field IEDs. It is has a hierarchical object-oriented data structure that enables the user to define data and attributes such as configuration information, naming, and diagnostic information. It can be used to browse and retrieve data from devices based on these attributes, without specific knowledge on the device's configurations or other details. The network stack of IEC 61850 is shown in Figure 6.3.

The performance requirements for various power grid applications vary quite widely – from low data rate with very low latency of less than 4 ms for protective relaying applications to SCADA reporting with low data rates and higher latencies allowed. 61850 accommodates these different applications by creating fields with different interfaces to the rest of the network stack, thus allowing prioritization of different kinds of messages. The time-critical messages such as protective relaying are mapped directly to Ethernet layer frames using non-IP protocols. These messages include the Sampled Measured Values (SMV/SV), the Generic Object Oriented Substation Events (GOOSE), the Generic Substation State Events (GSSE), and the Manufacturing Messaging Specification (MMS). MMS allows messages to be transferred directly through the Ethernet layer, or they can be transferred through TCP/IP connections depending on the application. The time synchronization (TimeSync) messages are transferred through UDP/IP connections.

61850 uses a different data naming convention, which isolates the power system application and object models from the communication protocols. This means

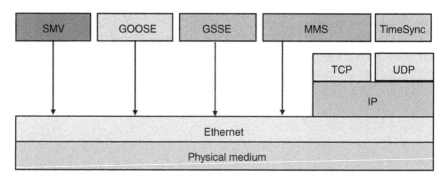

Figure 6.3 IEC 61850 network stack.

that there is no manual mapping needed between the power system functions of different devices from different vendors to the protocol objects, as this definition already exists within the protocol. This aspect of the data structure is explained more in Chapter 7.

6.3 ICCP

Multiple (in the order of thousands) business entities run different parts of the smart grid. These include generation companies, transmission system owners and operators, distribution system asset owners, and operators. The grid must always operate synchronously to ensure that the interconnections do not collapse. This requires coordinated decision making between the various control centers that operate various regions of the power grid. This coordinated decision making is enabled by information sharing between the control centers using the Inter-Control Center Communications Protocol (ICCP), which is part of the IEC 60870 protocol suite under the name Telecontrol Application Service Element 2 (TASE.2). ICCP is a client–server architecture-based protocol operating at the application layer, which used TCP for transport. The challenge with inter-control center communications is that the communication needs to be real time, with high availability and reliability in communication. This is achieved through the client–server architecture with performance guarantees through the use of TCP and other enhancements. A general use case for the ICCP is shown in Figure 6.4.

The ICCP protocol has several limitations in the smart grid area that is now under active research and development:

1. Lack of a uniform data management system – there is no coordination between various entities using the ICCP protocol on standard naming conventions that will avoid conflict. Currently, bilateral agreements need to be setup between partners to avoid conflicts.
2. The protocol might not be fast enough for some time-sensitive, real-time applications, as it relies on data that are polled from a historian that operates without high priority.
3. It uses regular TCP and needs enhancement to secure transport layer implementations such as TLS.

6.4 C37.118

The PMU, whose measurement aspects have been discussed in previous chapters, is a protocol to communicate synchrophasor data, which is significantly faster

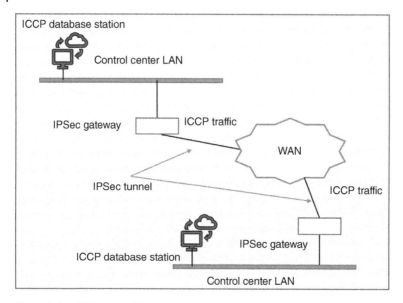

Figure 6.4 ICCP protocol between control centers.

than regular SCADA measurements. Synchrophasor measurements are generated at 30–240 times per second, and each measurement is taken under microsecond accuracy, which is significantly more accurate than SCADA measurements. PMU measurements are communicated using the IEEE C37.118 standard, and it is an application layer protocol that can use either UDP or TCP for the transport layer depending on the power system application requirements. The C37.118 protocol is intended to convent synchrophasor data from the field PMUs to the phasor data concentrator (PDC) and also to communicate and aggregate data between PDCs. The data format for the C37.118 protocol is shown in Figure 6.5.

The PMU protocol exemplifies the modern sensing paradigm that leverages the improvements in communication and computations technology. PMUs have a high reporting rate, are accurately time-stamped, and generate a large amount of data that requires large bandwidth and advanced computation capabilities to receive, process, analyze, and process the data. The combined improvement across the entire cyber-physical spectrum allows the grid to be much more automatic and move toward real-time control and coordination. PMUs provide better visibility and situational awareness to system operators and represent a huge improvement over just SCADA measurements. Currently, various grid applications are being developed and deployed by utilizing synchrophasor measurements that enable better reliability in power supply, greater efficiency in power delivery, and increased integration of highly variable renewable generation such as solar and wind technologies.

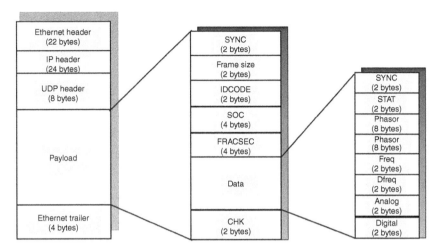

Figure 6.5   IEEE C37.118 data format.

The PMU technology is not without its drawbacks such as the following:

1. Lack of common data format for phasor measurements and problems with data sharing between utilities and vendors
2. Lack of flexibility in sampling and polling for measurements, as not all applications will require high-resolution information
3. Lack of methods and technologies to process PMU data into intelligent, actionable information for real-time control
4. PMU to PDC application path imposes latency on the system that might need improvement for faster grid applications

These and more problems are areas of active research in the smart grid area.

6.5   Smart Metering and Distributed Energy Resources

While PMUS represent the paradigm shift in the transmission system, distributed grid-edge intelligence is gaining traction at the distribution system level. The distributed intelligence can be grouped into two major categories – (i) smart metering systems and (ii) distributed energy resources (DERs).

6.5.1   Smart Metering

The power grid operator has historically not had much visibility at the grid edge, as they were reliant on SCADA measurements from the distribution system feeders, which were sparsely present from the substation to the consumer. In most cases, consumers had to call and inform the utility that the power had gone out in the case of any failures, as the utility had no visibility to the power flow to individual consumers. However, with smart metering systems, the utility is increasingly gaining more situational awareness at the grid edge. Smart meter deployment has largely been the focus of digitization of the distribution grid. An average of 65% of customers in North America, Europe, and Asia are already using smart meters, with a projected increase to 85% in the coming years. The benefit of smart meters for the utilities and system operators is that smart metering enables them to have better visibility of their networks and reduces operation and maintenance costs. It can be considered as a sensor, which can "open the door" to new services. Among the requirements for these smart meter technologies, there must be the following:

1. Two-way communication between the smart metering system and external networks for maintenance and control of the metering system;
2. Support of advanced tariff systems;
3. Provision of data for at least 15 minute granularity, where most EU member states currently support 60 minutes.

A general smart metering architecture is shown in Figure 6.6.

While the transmission system utilizes the wide-area network (WAN), smart metering technology relies on the home area network (HAN) and the neighborhood area network (NAN). HAN is the network inside a home or a building and can use a variety of communication protocols such as ZigBee, Z-Wave, Wi-Fi, or

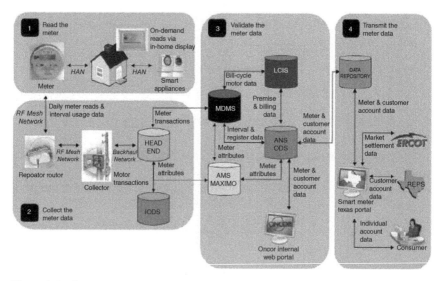

Figure 6.6 General smart metering architecture.

other home automation protocols. The rapid adoption of IoT and IoT networks is also contributing to the increase in automation and "smartness" at the grid edge. Specifically, utilities are expected to one of the highest users, with 1.37 billion endpoints. Gartner Inc.'s report also states that "electricity smart metering, both residential and commercial, will boost the adoption of IoT among utilities." Thus, HANs will be crucial to integrate not just the growing number of smart appliances and other IoT devices but also integrate them with utility endpoints such as meters and sensors. An evolving challenge is the need to standardize the protocols being used at this level, the data format to be used for interoperability, and methods for ensuring privacy and security of data.

On top of the HAN, NANs are crucial to coordinate the increasing number of smart meters and sensors at a community or aggregated level. Most protocols at this level are proprietary protocols, which reduce the scope for interoperability. There are also mesh routing technologies that utilize radio signals based on pole-top stations. The use of smart meters and integration of IoT devices into grid operation is an area of active research.

6.5.2   Distributed Energy Resources (DERs)

DERs include rooftop PV units, smaller wind turbines, electric vehicles, and customer flexibility in reducing power consumption that is also referred to as demand response. These technologies allow the user to be active participants in the grid operation instead of passive consumers. The regulation around the use of these DERs is still evolving, and there are similar problems as with smart metering – namely, the presence of multiple proprietary standards and protocols being used currently and lack of coordination and interoperability with the existing grid resources. While other protocols such as Modbus, IEC 61850, and even ZigBee have been deployed previously, a recent standard that has been gaining traction is the IEEE 2030.5 Smart Energy Profile (SEP) 2.0 standard, which provides a framework for monitoring and control of DER assets. It has been gaining traction with grid operators and has been suggested as the standardized communication protocol for DER aggregation programs. California's Rule 21, for example, describes various grid services that connected DERs should be able to perform, and the Common Smart Inverter Profile (CSIP) describes how SEP should be implemented to meet Rule 21. CSIP requires that DERs be capable of monitoring and reporting operational characteristics and perform grid support functions such as operating in a "Volt-Var control" mode. Regulatory policy is needed to dictate the limits of grid support functions that can be performed by DERs, and the ways of connecting these assets to the grid need to be standardized.

6.6   Time Synchronization

With the increasing number of sensors and control devices in the grid, time synchronization of measurements and various devices becomes important. The Global Positioning System (GPS)-based clocks are pretty cheap to source, but installation and interface of these devices to the grid and grid operations software is not trivial. There is a proposal to use the ubiquitous Ethernet to distribute time across various devices to create synchronization between them. There are two major protocols that can be used for time synchronization over Ethernet:

1. Network Time Protocol (NTP) RFC 5905 – This protocol is primarily intended for wide-area time synchronization. It is widely deployed across the Internet and has been in use for 35+ years. It has an accuracy of about 10 ms usually, although it can vary according to conditions.
2. Precision Time Protocol (PTP) IEEE 1588 – This protocol is primarily intended for local area time synchronization. It was first proposed in 2002 and is now gaining traction for adoption. It is capable of providing microsecond agreement on a LAN.

The basic idea of distributing time over Ethernet is as follows:

1. Host A sends a message and gets a reply from another host – now, Host A can calculate the round-trip time
2. Host A sends a message with its current time and the round-trip time to Host B
3. Host B can now estimate the offset between their local clocks, thus allowing the two hosts to be time synchronized

This approach is heavily dependent on the accuracy of their local clock sources. There are also additional problems because of the variability in communication due to network traffic and asymmetric communication times between steps 1 and 2. This is a manageable issue for a strictly controlled LAN where the propagation delay and other components in latency are small and can be estimated but becomes a bigger problem for WANs. Also, depending on the timing accuracy required, the time capture and time stamping need to happen at a low level (physical or link layer) of the communication stack, which poses a problem as these protocols are not widely deployed in the grid now. The question of time synchronization for WANs and for a large number of devices is an area of active research in the smart grid.

6.7   Summary

In this chapter, we discussed the typical power system application layer protocols and how they are different from common application layer protocols such as those used in the Internet. Two common SCADA protocols were discussed – DNP3 and

IEC 61850 – along with newer protocols such as the PMU protocol IEEE C37.118. In addition, application layer protocols that are evolving in the distribution system applications were discussed, such as the protocols used in smart metering or for DER resources. The time synchronization of these protocols with other devices in the grid has been briefly discussed.

6.8 Problems

1 What is HMI in a SCADA system?
   A Human–Machine Interface
   B Human–Machine Intelligence
   C Human-Managed Interface
   D Human-Managed IEDs
   E Human–Machine Interrupts

2 What does MMS stand for, in IEC 61850?
   A Multimedia Messaging System
   B Multimedia Messaging Service
   C Manufacturing Messaging Specification
   D Manufacturing Messaging Service
   E Multisystem Messaging Service

3 Which protocol is used to communicate between smart grid control centers?
   A DNP3
   B ICCP
   C HTML
   D IEEE C37.118
   E TCP/IP

4 California's Rule 21 specifies that DER's within Investor-Owned Utilities must utilize IEEE 2030.5 protocol as described in the CSIP. What does CSIP stand for?
   A California Smart Inverter Profile
   B California Solar Inverter Profile
   C Common Solar Inverter Profile
   D Common Smart Inverter Profile
   E Common Smart Inverter Protocol

5 What is typically the accuracy of the Network Time Protocol (NTP)?
   A 1 ms
   B 10 ms

C 100 ms
D 500 ms
E 1000 ms

6.9 Questions

(1) Explain the components and deployment of a typical SCADA system.
(2) What are the features of the DNP3 protocol? Are there any disadvantages of utilizing the DNP3 protocol for smart grid applications?
(3) How do smart grid control centers communicate with each other?
(4) Discuss the features of the PMU protocol, with its advantages and disadvantages.
(5) How is smart metering deployed? What is its network architecture and what protocols are typically used?

Further Reading

Antón, S.D., Fraunholz, D., Lipps, C. et al. (2017). Two decades of SCADA exploitation: a brief history. *2017 IEEE Conference on Application, Information and Network Security (AINS)*, pp. 98–104. https://doi.org/10.1109/AINS.2017.8270432.

Bani-Ahmed, A., Weber, L., Nasiri, A., and Hosseini, H. (2014). Microgrid communications: state of the art and future trends. *International Conference on Renewable Energy Research and Application (ICRERA)* 2014: 780–785. https://doi.org/10.1109/ICRERA.2014.7016491.

Finster, S. and Baumgart, I. (2014). Privacy-aware smart metering: a survey. *IEEE Communication Surveys and Tutorials* 16 (3): 1732–1745. https://doi.org/10.1109/SURV.2014.052914.00090.

IEEE Std C37.118.2-2011 (2011). *IEEE Standard for Synchrophasor Data Transfer for Power Systems* (Revision of IEEE Std C37.118-2005), pp. 1–53. https://doi.org/10.1109/IEEESTD.2011.6111222.

Mackiewicz, R.E. (2006). Overview of IEC 61850 and benefits. *IEEE PES Power Systems Conference and Exposition* 2006: 623–630. https://doi.org/10.1109/PSCE.2006.296392.

Mohagheghi, S., Stoupis, J., and Wang, Z. (2009). Communication protocols and networks for power systems-current status and future trends. *IEEE/PES Power Systems Conference and Exposition* 2009: 1–9. https://doi.org/10.1109/PSCE.2009.4840174.

Pliatsios, D., Sarigiannidis, P., Lagkas, T., and Sarigiannidis, A.G. (2020). A survey on SCADA systems: secure protocols, incidents, threats and tactics. *IEEE Communication Surveys and Tutorials* 22 (3): 1942–1976. https://doi.org/10.1109/COMST.2020.2987688.

7

Utility IT Infrastructures for Control Center and Fault-Tolerant Computing

In this chapter, we will understand the necessary infrastructure for a modern control center and how the requirements will evolve in the future. We will also study the data management in smart grids, looking at the challenges in interfacing devices and algorithms developed at different times.

7.1 Conventional Control Centers

The impetus to have control centers came from the 1965 Northeastern blackout, with increased government impetus to better monitor and control the power grid. To provide increased visibility in terms of grid stress, energy management systems (EMS) were created around the 1980s. These systems still used fairly crude Information and Communication Technology (ICT) systems, even by the 1980 standards. The power grid ICT systems have not been upgraded iteratively over the years, and by 2000, the technology was largely obsolete compared to modern technology. This situation has been improving since the turn of the century, with steady improvements being made. However, the ICT of the power grid is still lagging behind compared to modern technology.

In the 1950s, the power grid was becoming more connected, and measurement and control technology was being deployed. At this stage, analog measurements were communicated to analog computers to make control decisions. Grid applications such as load frequency control (LFC) and economic dispatch (ED) were common. Unit commitment (UC) was used to create schedules for start/stop of generators. As sensing and communication technology improved, remote terminal units (RTUs) were developed. The RTUs were used to measure signals at grid edge, which were then carried to a central EMS using Supervisory Control and Data Acquisition (SCADA). While SCADA and RTUs led to better visibility on the system, the information was not being used for real-time

Cyber Infrastructure for the Smart Electric Grid, First Edition.
Anurag K. Srivastava, Venkatesh Venkataramanan, and Carl Hauser.
© 2023 John Wiley & Sons Ltd. Published 2023 by John Wiley & Sons Ltd.

control. The 1965 Northeastern blackout led to several recommendations and technological advancements, including the advancements and adoption in phasor measurement technology.

One of the key advancements from the 1965 blackout was the use of digital computers (replacing the analog computers) to extensively monitor and control the power grid in real time. This advancement led to the concept of "power system security," which refers to the ability of the power system to withstand disturbances and contingencies. This also evolved into having a "$n - 1$" security constraint, by which the power grid has to be able to withstand the loss of any one component at any time. The development of digital computers also led to the replacement of special purpose computers for EMS with general purpose computers running EMS at the control centers. With reducing cost of technology for sensors and communication architecture, SCADA was also installed in distribution systems to create distribution management systems (DMS). The architecture of a conventional control center is shown in Figure 7.1.

With increased emphasis on the "security" of the power grid, Security-Constrained Economic Dispatch (SCED) was put into practice. Other grid services to support security were also created, including real power reserves and spinning reserves and ancillary services. Ancillary services included services other than real power such as voltage support, load following, and so on and were often traded on an inter-utility basis. Till this time, utilities were vertically

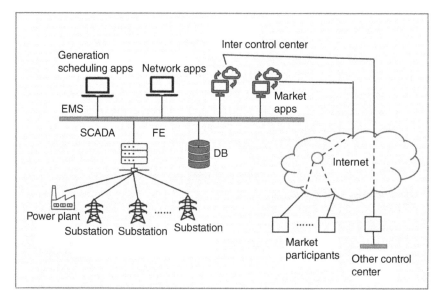

Figure 7.1 Architecture of a conventional control center.

integrated, with a single entity responsible for generation, transmission, and distribution of power. In the mid-1990s, the electricity industry was deregulated, and the three components of the power grid were separated. This instituted a number of changes, chief among which was the creation of Independent System Operators (ISOs) or Regional Transmission Operators (RTOs) who were responsible for enforcing the standards on the transmission of the power from the generators to the Load Serving Entities (LSEs). A two-level structure evolved at the control center, with the reliability/security focused on EMS and an economic/business functionality to optimize the price of electricity. There exists two types of markets – (i) *bilateral contracts* market, between suppliers and consumers, and (ii) *auction market*, where generators submit bids to supply energy at a particular cost. There also exist markets for day-ahead bids, real-time balancing (approximately 5 minutes), and ancillary services.

The National Energy Reliability Corporation (NERC), an independent body, was formed to be responsible for reliability in the power grid. They act as a balancing authority (BA) and ensure reliability during normal steady-state operation.

7.2 Modern Control Centers

Modern control centers are the result of cumulative improvements made to the grid, communication, and computation architectures made over time. An example is the various Intelligent Electronic Devices (IEDs) that are used in substations. These IEDs provide lots of useful data used by different applications in the power grid. This also provides a challenge, as the sensors/IEDs have been developed at different times and hence use various proprietary protocols and technologies. These data have to be collected from different points in the grid, various copies to be made, and must be coordinated, synchronized, and merged into usable databases.

Figure 7.2 shows a modern control center architecture. Substations have LANs connecting IEDs to RTUs. These can be simply Ethernet connections or have other power system protocols such as IEC 61850. The challenge with the SCADA architecture is that it was developed when utilities were vertically integrated, and the structure still remains the same. Hence, coordinating the data between various RTUs and power system applications provides a challenge. However, the scenario is changing in recent times, with the vertically oriented utilities tending toward deregulation. Control centers often supervise non-contiguous geographical areas based on mergers, acquisitions, and divestitures between utilities. The hierarchical structure of EMS, SCADA, and DMS are being challenged by the advent of grid-edge sensing and computation. Hence, we need flexibility in

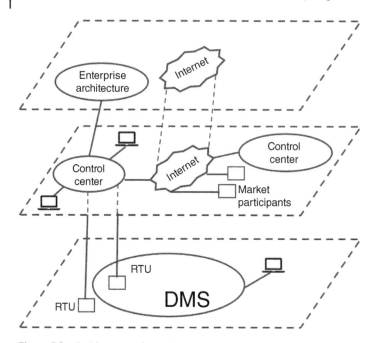

Figure 7.2 Architecture of a modern control center.

many dimensions and at many levels to incorporate modularity, scalability, and expandability toward a coherent enterprise architecture.

The solution toward a coherent architecture is by utilizing software expertise from other domains and tools such as Enterprise Resource Planning (ERP), also known as Enterprise Resource Management (ERM). ERP manages all aspects of a business such as production planning, material purchasing, maintaining inventories, supplier interactions, payroll, and customer service. Figure 8.1 shows a reference architecture of modern control centers integrating with various other players horizontally and also vertically. Various initiatives such as the North American Synchrophasor Initiative (NASPI) are facilitating this transition.

7.3 Future Control Centers

Future control centers will be a system where the power system applications and services are spread out on multiple computers at different locations or on the cloud. These computers will only interact via message passing, with no shared memory. This *decentralized* architecture will enable better redundancy and fault tolerance in the system. Various control architectures will be pushed to the substation level, such as special protection schemes (SPS) or remedial action schemes (RAS). A common data interchange format, such as the Common Information

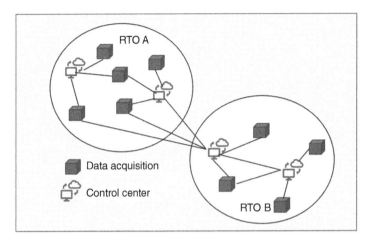

Figure 7.3 Architecture of future control centers.

Model (CIM), will become more prevalent. Middleware-based EMS and business functions will become common to enable better coordination. Data acquisition will be at a much higher rate and at a higher amount. The existing grid sensors such as Digital Fault Recorders (DFRs) already report at a very high rate of 720+ Hz. Future sensors might report at a rate as high as 5000 Hz. An example of such distributed data acquisition is shown in Figure 7.3. This faster data acquisition does not need to be the norm for all grid sensors, as this depends on the type of application that uses this data. Data acquisition rates will depend on the latency required for the application, criticality of the application, quantity of the data, geographical distance shared over, and degree of sharing (number of subscribers) per sensor.

The future control centers can only be possible through the following:

1. Cleanly decomposed and packaged application logic
2. Separate management of data acquisition, managing computational resources, information security, etc.
3. Cloud computing for power grid services.

This clean separation enables new technology (both apps and ICT) to develop much faster, which worked for the internet. The Quality of Service provided by the communication technology still presents a bottleneck for future development.

7.4 UML, XML, RDF, and CIM

From the description of the control centers above, it is clear that moving data from distributed locations to a central location and exchange of data between neighboring devices will play a critical role in advancing control center architectures. To enable this robust data sharing, it is important to consider the format of the data that is being transferred.

EMS has typically had application-specific, proprietary file formats that are often crude: column oriented, fixed width, and tab separated, which is similar to FORTRAN. Deregulation of markets is pushing vendors, utilities, and other stakeholders from doing their own custom software and data formats to ensure that their products have a wide adoption across the domain. Hence, interoperability is becoming key in the evolving EMS world. However, problems arise when sharing data between different vendors/utilities' software and sharing data even between two versions of same software. To address this compatibility issue, several solution methodologies are possible:

1. Maintain multiple copies of the same data in multiple formats
2. Store the data in a format compatible with every piece of software and hence have to remove app-specific data and lose precision
3. Store data in a single, highly detailed format and create software to convert to app-specific formats
4. Use a highly detailed format compatible with every application

Of these, #4 is the ideal solution as it does not require storing multiple copies and reduces the need for conversion between various applications. To create such a detailed format, there are a few requirements to be satisfied:

1. Highly detailed model to describe the power system
2. File format capable of storing extended data without affecting the core data
3. Power system software vendors and utilities adopt and embrace the data model

Some solutions that address requirements 1 and 2 include the CIM, Extensible Markup Language (XML), Unified Modeling Language (UML), Resource Description Framework (RDF), and certain power system frameworks such as IEC 61850.

7.4.1 UML

The Unified Markup Language (UML) is used for modeling a wide variety of software components such as (i) data structures, (ii) system interactions, and (iii) use cases. The advantage of UML is that it is not tied to any one implementation technology and is realizable on multiple platforms. UML is used widely outside the electricity sector and is standardized by the Object Management Group (OMG). OMG is an organization that largely works on middleware platforms and creates frameworks and standards.

UML utilizes concepts from Object-Oriented (OO) programming such as Class, Inheritance, Association, Aggregation, and Composition.

Circle
X
Y
Radius

Figure 7.4 UML class.

Class represents a specific type of "object" being modeled. Class hierarchy is an abstract model and defines every type of component within a system as a separate class. An UML example is shown in Figure 7.4.

Inheritance defines class and sub-class relationships where sub-classes inherit attributes from parent class (and up). In UML as shown in Figure 7.5, inheritance is represented by a clear triangle on the parent side of the line.

Associations are used to describe relationships other than parent–child, which may be useful. An UML example is shown in Figure 7.6.

Composition is a specialized form of aggregation where the contained object is a fundamental part of the container object. In the example, an **Anchor** class shows where a line can be attached (anchored) on a **shape**. The UML depiction is a filled diamond on the container end, as shown in Figure 7.7.

Aggregation is a special kind of association, indicating one is a container of instances of the other. The UML depiction is a line with a clear diamond on the

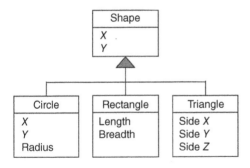

Figure 7.5 UML inheritance

Figure 7.6 UML association.

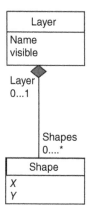

Figure 7.7 UML composition.

Figure 7.8 UML aggregation.

container end, and with role and cardinality defined at each end. In the example shown in Figure 7.8, a **layer** that contains **shape** objects and can be toggled on and off.

7.4.2 XML and RDF

XML is a markup language based on SGML (Standard Generalized Markup Language), like HTML (HyperText Markup Language), which is the most basic building block of webpages. XML works by defining a set of rules for encoding documents/data that is both human readable and machine readable. XML's design goals are simplicity, generality, and usability over the Internet. Elements expressed in XML are expressed as

```
<tag> … contained data … </tag>
```

An example of a XML specification for a book is as follows -

```
<book title="Hi" author="Joe">
<rev num="2">
<year>2006</year>
<month>January</month>
<day>1</day>
```

```
</rev>
<chapter title="Preface">
<paragraph> .... </paragraph>
....
</chapter>
.... (another chapter)
</book>
```

It is important to note that XML has no set syntax or semantics for a tag. In the context of a smart grid, power system applications have to know the tag and be able to parse them to interpret an XML document. This is achieved by the use of a XML schema. XML Schema Definition (XSD) defines (i) elements and attributes that may appear in a message, (ii) parent–child relationships, (iii) number of children allowed for an element, (iv) whether an element can include text, (v) data types for elements and attributes, and (vi) whether data items have fixed values or default values.

The RDF is an extension of the XML format. XML only has parent–child links. However, RDF improves this by allowing for generalization of relationships. For example, a **book** object can have **sequel** and **sequelto** attributes that allow for broader definition of relationships.

7.4.3 CIM (IEC 6170)

CIM is an implementation agnostic model for defining power grid data in an UML environment used by electric utilities. For example, a circuit breaker object has the following components:

1. **IdentifiedObject**: root class
2. **PowerSystemResource**: any resource in a grid
3. **Equipment**: a physical device (electrical or mechanical)
4. **ConductingEquipment**: equipment that carry current or are electrically connected to the grid
5. **Switch**: device that opens and closes
6. **ProtectedSwitch**: operated by protection equipment
7. **Breaker**: Specific breaker ID

This structure is shown in Figure 7.9.

7.4.4 IEC 61850

IEC 61850 is fast becoming the de-facto standard of data models for smart grid communications. IEC 61850 was designed keeping in mind the unique requirements for smart communications and has several distinguishing features including the following:

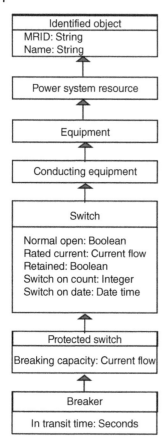

Figure 7.9   CIM for a breaker.

1. *Data modeling*: The data modeling for 61850 is elegant and has primary process objects as well as protection and control functionality that are ubiquitous in substations. They are modeled into different standard logical nodes, which can be grouped under different logical devices.
2. *Reporting schemes*: The standard allows for server–client relationships on which communication can be triggered both on schedule and by pre-defined trigger conditions.
3. *Fast transfer of events*: The model defines Generic Substation Events (GSE) that allows fast transfer of information. It is further divided into GOOSE (Generic Object Oriented Substation Events) and GSSE (Generic Substation State Events).
4. *Data storage*: Substation Configuration Language (SCL) is defined for complete storage of configured data of the substation in a specific format.

The IEC 61850 standard came about due to the increasing fragmentation and proprietary protocols and data formats that were being used by the industry. The objectives on which the standard was created is as follows (CITE STANDARD):

1. A single protocol for complete substation considering modeling of different data required for substation
2. Definition of basic services required to transfer data so that the entire mapping to communication protocol can be made future proof
3. Promotion of high interoperability between systems from different vendors
4. A common method/format for storing complete data
5. Define complete testing required for the equipment that conforms to the standard

The advantage of IEC 61850 is that it largely achieves the objectives that went into its design. The data model and SCL model are elegant and offer substantial advantages over legacy and current practices. However, it has also faced criticism such as follows:

1. It is far more complex than it has to be given the problem it is tackling
2. Double the size/bandwidth of IEEE C37.118 with no extra useful info
3. Subscriber apps have to be able to detect missing and duplicates (no sophisticated fault-tolerant multicast)
4. Shared-key multicast authentication flavor allows subscribers to spoof a publisher
5. Have to be careful that the multicast (Ethernet broadcast) does not overload small embedded devices, which are still a part of the legacy smart grid
6. 61850 assumes that the same interface for a LAN will work in a WAN, which might always hold true. However, this is partially addressed in IEC 61850-90-5, which is the WAN extension of the standard.

While IEC 61850 is fast becoming the most accepted standard, there is still work to be done to ensure that critical applications are well tested, and transitions are appropriately managed.

7.5 Basics of Fault-Tolerant Computing

Because the power grid is a critical infrastructure, it is essential that it has the capability to perform to its specifications even with disruptions to its operation. This introduces the need for fault-tolerant computing, which is defined as a system that can provide service despite one or more faults occurring. As the power grid is a dynamic system with potential for multiple failures, it cannot be fault resistant and hence has to be fault tolerant. In the fault-tolerant computing realm, it is important to familiarize with a couple of definitions:

1. Dependability is the measure in which reliance can justifiably be placed on the service delivered by a system. In addressing dependability, it is important to understand the impairments or threats to dependability, know the means to achieve it, devise ways of specifying and expressing the level of performance required, and measure performance to understand if dependability has been achieved.

2. Fault is the adjudged or hypothesized cause for an error. An important distinction to note here is that faults can lie dormant in a system for some time. An example is a software bug that might not occur in regular cases but is triggered by a niche case.

3. Error is an incorrect system state. An example is incorrect information given on the number of bytes on a disk for a given record.

4. Failure is the state when the component no longer meets its specification, i.e. the problem is visible outside the component and has an impact on performance.

The overall sequence of events is Fault → Error → Failure. Faults, if not handled properly, can cascade. An example of this is shown in Figure 7.10, where a fault in component 1 leads to an error, which results in a failure in component 1. Component 2 relies on the input from component 1, and hence, now component 2 experiences a fault, and this sequence can potentially cascade. There are several ways of classifying faults:

1. *Phenomenological origin*: these might be physical (hardware causes), caused by design, or occurring due to interactions between components
2. *Nature of faults*: they could be accidental or malicious
3. *Phase of creation*: faults could occur during development or operations
4. *Locus*: internal or external propagation
5. *Persistence*: faults could be temporary or permanent

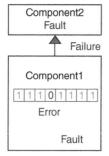

Figure 7.10 Cascading faults.

A non-exhaustive taxonomy of faults is presented in Table 7.1.

To achieve dependability in a system, it is important to break the chain of cascading faults. There are various techniques geared toward this:

1. *Fault removal*: detecting and removing faults before they can cause an error
2. *Fault forecasting*: estimating the probability of faults occurring or remaining in the system
3. *Fault prevention*: eliminate conditions that make fault occurrence probable
4. *Fault avoidance*: fault prevention + fault removal

Once the fault has been dealt with, it is important to quantify the performance of the system to ensure that it is within specification. Typically, a measure such as Mean Time Between Failures (MTBF) is used in fault-tolerant systems. In the power grid, there are several such measures such as SAIDI (System Average Interruption Duration Index) or SAIFI (System Average Interruption Frequency Index).

7.6 Cloud Computing

Cloud computing is fast gaining traction in many domains and has modernized many business enterprises. It is globally seen as a critical infrastructure like other vital resources such as power, gas, and freshwater supply. Cloud technology has major benefits over traditional, in-site forms of computation that often run into a problem with scale. However, the adoption of cloud technology in the power industry faces resistance on several fronts such as cybersecurity concerns, compliance with operational standards, cost of deployment, consistency in performance, latency in critical applications, and a general resistance to change. Table 7.2 compares the advantage of cloud technology over traditional on-site computation based on a report from an IEEE Taskforce Zhang et al. [2022].

In the power grid, the cloud hardware (big data centers with flexible and seemingly infinite computational resources) might have to be hosted by a government or stakeholder entity such as Department of Energy, NERC, or an ISO, although private players such as Microsoft and Amazon might also participate. Cloud computing allows for better consolidation of resources, with 10x or better reductions in cost of operation, with better equipment utilization and management, newer styles of elastic computing, and potential to compute directly on "big data." All of this adds up to a new way of computing that forces us to undertake new kinds of thinking. Cloud computing does come with a trade-off in consistency for its scalability, so that needs to be considered in all applications going forward. Cloud computing could be advantageous for utilities because of the following:

1. They are understaffed on the IT front,
2. Have an opportunity for much broader sharing of operational and planning data, and
3. Have an opportunity for very advanced data analytics.

Table 7.1 Taxonomy of faults.

| Nature | | Phenomenological cause | | Origin | | | | | | Usual labeling |
Accidental faults	Intentional faults	Physical faults	Human-made faults	Internal faults	External faults	Design faults	Operational faults	Permanent faults	Temporary faults	
				System boundaries		Phase of creation		Pesistence		
X		X		X			X	X		Physical faults
X		X			X		X	X		
X		X			X		X		X	Transient faults
X		X		X			X		X	Intermittent faults
X			X	X		X		X		
X			X	X		X		X		Design faults
X			X		X		X		X	
	X		X	X		X		X	X	Malicious logic
	X		X	X		X			X	
	X		X		X		X	X	X	
	X		X		X		X	X	X	Intrusions

Table 7.2 Advantages of cloud computing over traditional on-site computation.

Advantage	How is it used?
Agility	Quickly edit/create new computing infrastructure
Cost savings	Less capital investment in infrastructure, pay only for usage
Resilience	Highly available, able to recover from failures faster
Elasticity	Easily scale up/down resources per requirement
Support	Dedicated cloud support relives burden from users for troubleshooting

Cloud computing can be deployed in multiple models such as private cloud, public cloud, community cloud (restricted access to specific participants), or hybrid cloud. There are also various service models such as follows:

1. *Infrastructure as a service (IaaS)*: Lets consumer provision processing, storage, network, data, operating system, etc. This model allows the user to control resources at a low level.
2. *Platform-as-a-service (Paas)*: a category of cloud computing services that allows customers to provision, instantiate, and use a customized bundle of resources comprising a computing platform and any applications without users performing any essential management.
3. *Software-as-a-service (SaaS)*: a category of cloud computing services where a vendor hosts applications and delivers them to end users over the Internet on a subscription basis.

While a community-restricted cloud computing environment with either a IaaS or PaaS model that avoids problems such as vendor or data lock-in might be the best solution for the power grid, it remains to be seen how cloud computing is adapted in the future.

7.7 Summary

In this chapter, we have studied the evolution of control centers from conventional to modern. We discussed the computational requirements for the evolving smart grid, and the data management standards evolving around them, such as IEC 61850. We have also looked at the evolving need for fault-tolerant computing, its basics, and its potential future deployment on the cloud.

7.8 Problems

1 Find a false statement
   A The 1965 Northeastern blackout report recommended more extensive use of digital computers in real time
   B Security-constrained economic dispatch began after the 2003 blackouts in North American and Europe
   C Cloud computing offers great cost advantages
   D Bilateral contracts are between suppliers and consumers

2 Find a false statement about UML
   A An abstract class cannot be instantiated (have instances of)
   B Association is a relationship other than parent–child
   C UML is widely used outside the electricity sector
   D UML is realizable only for one implementation technology, so its models cannot be used

3 Find a false statement
   A XML's design goals include simplicity, generality, and usability over the Internet
   B XML is widely considered to be more readable to humans than UML
   C XML Schema Definition (XSD) defines the number of children allowed for an element
   D XML per se has no set syntax or semantics for a tag

4 Find a true statement
   A CIM is an implementation-agnostic model for defining data used by electric utilities
   B The equipment class in the CIM is the root class
   C The IdentifiedObject is a subclass (or sub-sub-class, etc.) of the equipment class
   D The breaker class is the parent class of the PowerSystemResource class

5 Find a true statement
   A The chain of cascading within a module is error –> failure –> fault
   B An error is when a component no longer meets is specification
   C A fault may lie dormant for some time
   D A fault in one component always directly and immediately leads to a fault in any component that uses it

7.9 Questions

(1) Discuss the architecture of conventional and modern control center and how they have evolved.
(2) Discuss the requirements from a future control center and their expected architecture.
(3) Describe the Common Information Model (CIM) and explain the CIM for a capacitor bank.
(4) How do you deal with cascading faults? Explain the steps involved.
(5) Explain cloud computing and how cloud computing can be adapted for the power grid and its potential advantages.

Further Reading

Bakken, D. (ed.) (2014). *Smart Grids: Clouds, Communications, Open Source, and Automation*. CRC Press.

Bera, S., Misra, S., and Rodrigues, J.J.P.C. (2015). Cloud computing applications for smart grid: a survey. *IEEE Transactions on Parallel and Distributed Systems* 26 (5): 1477–1494. https://doi.org/10.1109/TPDS.2014.2321378.

Coulouris, G.F., Dollimore, J., and Kindberg, T. (2005). *Distributed Systems: Concepts and Design*. Pearson Education.

EPRI (2022). Common Information Model Primer: Eighth Edition. https://www.epri .com/research/products/000000003002006001 (accessed 29 August 2022) [Readers are advised to consult the latest version of the primer as released from EPRI].

Fox, A., Griffith, R., Joseph, A.D. et al. (2009). Above the clouds: a Berkeley view of cloud computing. Dept. Electrical Eng. and Comput. Sciences, University of California, Berkeley, *Rep. UCB/EECS* 28 (13): 2009.

IEEE Smart Grid Vision for Computing: 2030 and Beyond Roadmap (2016). IEEE Smart Grid Vision for Computing: 2030 and Beyond Roadmap, pp. 1–14.

Verissimo, P. and Rodrigues, L. (2001). *Distributed Systems for System Architects*, vol. 1. Springer Science & Business Media.

Wu, F.F., Moslehi, K., and Bose, A. (2005). Power system control centers: past, present, and future. *Proceedings of the IEEE* 93 (11): 1890–1908. https://doi.org/10 .1109/JPROC.2005.857499.

Zhang, S., Pandey, A., Luo, X. et al. (2022). Practical adoption of cloud computing in power systems—drivers, challenges, guidance, and real-world use cases. *IEEE Transactions on Smart Grid* 13 (3): 2390–2411.

8

Basic Security Concepts, Cryptographic Protocols, and Access Control

8.1 Introduction

This chapter explores the basic concepts in cybersecurity and then introduces many of the foundational security mechanisms used to protect power systems. A variety of security controls are necessary to protect against the numerous ways that an attack could manipulate the system. Furthermore, there are significant challenges in the implementation of these controls within constrained environments, such as substations that are geographically disperse and have limited computational resources. The chapter will introduce basic concepts in encryption, authentication, and access control while also exploring a variety of challenges implementing these controls. First, this chapter will introduce basic terminology and concepts within cybersecurity.

8.2 Basic Cybersecurity Concepts and Threats to Power Systems

Achieving perfect security is not possible; therefore, it is important to understand the key threats and security goals in order to identify an appropriate security solution. This section will explore a number of fundamental properties within cybersecurity and define a number of basic terms. Furthermore, it will provide power system examples of each of these concepts.

8.2.1 Threats, Vulnerabilities, and Risks, What Is the Difference?

Securing a system depends primarily on understanding its vulnerabilities, the external threats to the system, and the risk they impose. The remainder of this section will introduce the concepts of threats, vulnerabilities, and risks while also exploring how they relate to modern power grids.

Cyber Infrastructure for the Smart Electric Grid, First Edition.
Anurag K. Srivastava, Venkatesh Venkataramanan, and Carl Hauser.
© 2023 John Wiley & Sons Ltd. Published 2023 by John Wiley & Sons Ltd.

8.2.2 Threats

A threat is defined as a circumstance or an event with the potential to adversely impact organizational operations. Within the context of the power grid, a threat is any event that could adversely affect consumers, utilities, or the systems responsible for the flow of electricity. While threats can arise from a variety of domains (e.g. weather, mechanical failures, and wildlife), this chapter will primarily focus on cybersecurity threats, which introduce many new challenges to protecting the system. Cybersecurity threats typically can include (i) malicious or inadvertent functions by people, (ii) software vulnerabilities, (iii) system misconfigurations, (iv) cryptographic flaws, and (v) failed access control or authentication.

Recently, many sophisticated organizations or nation states have spent considerable efforts developing offensive cybersecurity capabilities. These capabilities are often described as advanced persistent threats (APT), defined as an attacker that (i) has a sufficiently large budget to fund professional hackers, (ii) sufficient persistence to continually look for weaknesses in a target system, and (iii) the ability to utilize sophisticated malware based on zero-day vulnerabilities and rootkits. These three factors make APT's extremely difficult to protect against.

The grid has many engineered redundancies, such as $N-1$ contingency, so many threats are unlikely to cause a serious interruption. Credible threats to the grid will likely be the result of coordinated attacks, defined as multiple simultaneous attacks to multiple parts of the grid, in order to bypass the redundancies and cause a significant interruption. A coordinated attack would likely have to both manipulate the situational awareness capabilities of utilities and reliability coordinators while also causing some instability to the bulk power system.

8.2.3 Vulnerabilities

A vulnerability is defined as a weakness or flaw in the system that could be exploited by a threat to cause some adverse effects to a system. Therefore, in order for a threat to be credible, it must have a vulnerability, or weakness, to exploit. Within a cybersecurity context, there are a variety of potential areas where system vulnerabilities could arise, including the networks, systems, and software.

Network: There are many different types of networks used to support the power grid; however, many of these may not be fully trusted, especially over a wide-area network. Data sent across a network may be vulnerable to eavesdropping, tampering, or spoofing. Furthermore, networks are also heavily dependent on a variety of support and management protocols, whose messages could also be spoofed to manipulate the operation of the network. A variety of encryption and authentication techniques are typically deployed to protect networks.

Systems: The various devices used to monitor and control the grid can also have a variety of vulnerabilities. These devices must communicate with a variety of different systems, some of which may be malicious or have untrusted users. Therefore, the systems must be able to verify the identity of other systems while also limiting the amount of information that system can access. However, incorrect implementations of the authentication and access control functions can provide individuals with access to unauthorized data or functions.

Software: Modern software is extremely complex. Many situations arise where software may not sufficiently handle incoming messages or user input, which could cause it to improperly management memory, overwrite critical control data, and allow an attacker to usurp its operations. This could ultimately provide an opportunity for an attack to remotely access a system or gain additional privileges on that system. There have been a variety of key software vulnerabilities in commonly used software devices and platforms.

8.2.4 Risk

The overall risk to the system depends on the combination of vulnerabilities, threats, and their impacts or consequence to the system. Threats to the system need to have some available vulnerability to exploit before they can adversely impact the system. This concept can be viewed in Figure 8.1 from

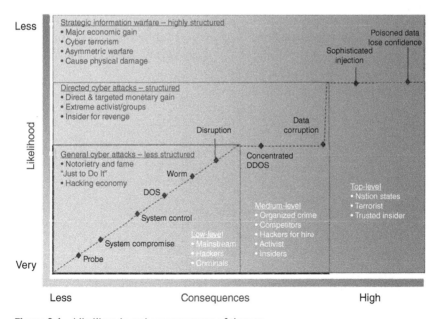

Figure 8.1 Likelihoods and consequences of threats.

NERC, exploring the likelihood and consequence of various power grid threats. The figure suggests that there are many probable attacks that might have limited system compromising or Denial of Service (DoS) event but are performed by low-level attackers and therefore have a limited impact on the system. On the end of the spectrum, unlikely attacks from top-level nation state threats could likely have severe consequences if they are able to perform significant injections of attack; however, these attacks have very low likelihood.

8.3   CIA Triad and Other Core Security Properties

The field of computer security is traditionally defined as the effort to provide confidentiality, integrity, and availability (CIA) to the information systems and data used to perform some tasks. Their key properties are defined by NIST FIPS-199 as follows:

1. *Confidentiality*: Preserving authorized restrictions on information access and disclosure, including means for protecting personal privacy and proprietary information.
2. *Integrity*: Guarding against improper information modification or destruction and includes ensuring information non-repudiation and authenticity.
3. *Availability*: Ensuring timely and reliable access to and use of information.

While computer security has traditionally emphasized confidentiality as the key factor, in many cyber-physical systems, such as the power grid, integrity and availability are considered the most critical properties. To elaborate this, we explore an example of Supervisory Control and Data Acquisition (SCADA) server and define the CIA requirements in the following table:

Property	Importance	Explanation
Confidentiality	Low	The data stored on this system, such as the voltage and current of lines or the status of breakers. If an attacker is able to gain access to this data, it alone has little value to them as they cannot directly use it to perform any additional action
Integrity	High	The reason for this is that an attacker could send circuit breaker commands to limit electricity to certain power lines and ultimately cause an interruption
Availability	High	The system functions and data must be available to the operator in the case of a grid interruption or other event so that the operator can take the necessary recovery action. If the system is unavailable, it is likely that there is an interruption

In addition to the CIA triad, there are a number of additional security properties that should be introduced, including accountability, authenticity, and privacy.

Accountability: This property focuses on the ability to ensure that an action can be directly attributed back to some entity. For example, accountability is maintained if we use a digital signature with a message to verify that it comes from the correct sender. Furthermore, this would also prevent a user from repudiating the sending of that message or action.

Authenticity: The authenticity of the messages means that we can verify the source of the message or action. We defined authentication as the process of verifying authenticity.

Privacy: It is defined as user's or organization's ability to control when, how, and to what extent information about themselves will be collected, used, and shared with others. While privacy is related to confidentiality, the key difference is that confidentiality looks at the broad protection of information, and privacy focuses on the user's ability to control how their information is used.

8.3.1 Privacy and Consumer Data

Before AMI and smart meters, privacy was not considered a major concern in power systems. Utilities would record and collect periodic measurements regarding consumer energy usages, but the low-frequency collection could not be used to infer much about the individual consumer behaviors.

However, the incorporation of modern smart meters allows utilities to collect home energy consumption at must faster rates. These will often collect home energy consumption measurements from every 15–60 minutes. While these increased meter readings enable the implementation of demand response and time-of-use rates, they present challenges with privacy of the consumers as individual home loads may begin to get increasingly observable.

A good example of this is demonstrated in Figure 8.2, which demonstrates the ability to infer a household's activities by only measuring the overall energy usage because of the well-known load profiles. The data from this example demonstrates collection from a very high frequency, and therefore, this example overstates the utilities' ability to infer household behavior from current smart meter deployments.

8.3.2 Encryption and Authentication

Encryption is used to enforce the confidentiality of data, especially when sent over an untrusted network. Much of the communications within the power grid communicate operational values (e.g. SCADA), which do not have high confidentiality requirements. For example, typical SCADA values such as bus voltages and

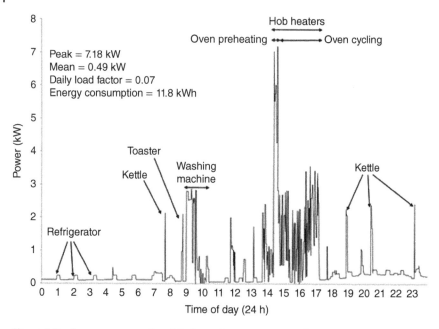

Figure 8.2 Energy consumption data impacts on consumer privacy.

current measurements do not have a significant value for an attacker. On the other hand, remote maintenance sessions to the substation may include information such as passwords and authentication tokens, which could potentially be reused by an attacker to gain access to the substation. Therefore, this information will almost certainly require encryption to protect the system from attack. Before we discuss the encryption techniques, we will first introduce some fundamental principles.

A basic cryptosystem includes an encryption algorithm $E()$ that accepts both a key, k, and a plaintext message, m. This function can be defined as $E(k, m) = c$, where c is the encrypted ciphertext that is likely to be tampered with by an attacker. There are two key approaches to perform encryption, symmetric key and asymmetric key, both of which have unique benefits and drawbacks. Figure 8.3 provides a basic overview of this process.

8.3.2.1 Kerckhoffs's versus Kirchoff's Law (Fundamental Cryptographic Principles and Threats)

There are many threats that must be considered with the application of any cryptosystem. The software necessary to implement cryptographic algorithms is extremely complex, and minor mistakes can completely degrade the security of the entire system. These threats can range from the theoretic and algorithmic

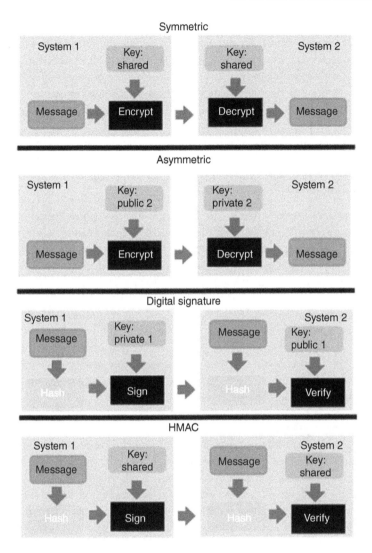

Figure 8.3 Symmetric and asymmetric encryption.

attacks to weaknesses in the implementation of the algorithms, including software vulnerabilities. The key threats include the following:

1. *Brute force attack*: A brute force attack attempts to enumerate all possible keys until it finds the correct one. This attack is often feasible on dated cryptographic algorithms with undersized keys; however, it is not commonly on modern ciphers.

2. *Cryptanalysis*: This is the process of finding pattern and correlations between plaintexts and the resulting ciphertext that allow the attacker to find either the key or plaintext from the ciphertext. This is typically the result of an algorithmic weakness in the encryption process.
3. *Software vulnerability*: As defined in an earlier chapter, software vulnerabilities are extremely prevalent because of the complexity of implementing the algorithms.
4. *Insufficient entropy*: May cryptographic techniques required random number generators to create keys and other values in the cryptographic protocols. If these values are not sufficiently random, the attacker may be able to infer either the key or the plaintext.

The development and implementation of cryptographic algorithms introduce many challenges and should only be performed by experts because of the severity of small failures. While electrical engineers learn Kirchoff's law as a fundamental principle, cryptography has Kerckhoffs's law, which states that a cryptographic algorithm should be publicly available and that only the key should be protected. The intent is that there is no security through obscurity and that it is far more secure to utilize industry standard cryptographic algorithms rather than the ones that are built with some proprietary ideas or techniques.

Warning: Because of a large number of possible errors that could degrade the effectiveness of the cryptographic protections. Therefore, any actual implementation of these techniques should be done with a professional. There are many caveats and nuances that cannot be addressed in this chapter.

8.3.2.2 Symmetric Key Encryption

Symmetric key encryption utilizes the same key to perform the encryption and decryption of a plaintext. This requires that the same key must be possessed by both the sender and the receiver, which introduces challenges on how the key exchange is performed. In this case, the encryption process is typically defined as $E(k, m) = c$, while the decryption algorithm is the inverse of the encryption algorithm, $E^{-1}(k, c) = m$. There are two different block ciphers approaches, stream ciphers and block ciphers. Currently, block ciphers, such as Advanced Encryption Standard (AES), are more popular, although certain stream ciphers are gaining popularity.

AES is currently the de-facto standard symmetric key block cipher. The cipher uses either 128, 192, or 256 bit blocks that are expected to provide strong security for many years to come. Furthermore, AES can be easily implemented in hardware, which means it can operate extremely quickly, typically in the order of microseconds. Therefore, AES is commonly used throughout the power grid.

Mode-of-operation: Because block cipher encryption is an n-bit block at a time, there may be times where similar plaintexts are encrypted to the same ciphertext.

While this is often not problematic, it may provide some weaknesses. Therefore, some mechanism is necessary to ensure that similar plaintexts are always decrypted, while there are many common modes-of-operation (e.g. ECB and CBC); the current ones are recommended that provide authenticated encryption (e.g. CCM, OCB, and GCM).

8.3.2.3 Asymmetric Key

Asymmetric key encryption differs from symmetric key encryption in that it utilizes two keys, public and private keys, which are unique for every person or device. The public key should be freely distributed to anyone and should be possessed by the individual attempting to communicate with each other; even potentially attackers can safely possess a public key without a security threat. The private key on the other hand must be protected by the owner and should never be disclosed to even trusted entities.

1. *RSA (Rivest, Shamir, Adleman)*: The RSA algorithm can be used to perform limited encryption functions. It can also be used to perform authentication functions through digital signatures. The RSA algorithm works on the difficulty of factoring large numbers.
2. *Diffe–Hellman*: Diffie–Hellman is a key exchange algorithm that allows two individuals to agree on the same key. This is not directly used to perform either encryption or authentication.

Elliptic Curve Cryptography Traditional asymmetric key cryptography occurs over the set of integers; however, this introduces challenges as the size of the keys must grow as the processor speeds improve. As on 2017, the standard for both RSA and DH is 2048 bits, which includes a significant overhead as devices must perform expensive operations over large exponentials and moduli. Fortunately, many of these same functions can also be performed over an elliptic curve (e.g. $y^2 = x^3 + ax + b$). Because efficient algorithms for factoring and computing discrete logarithms are not known, ECC operations can be performed with smaller keys and improved performance. For example, 224-bit ECC DHE is roughly equivalent to the 2048-bit traditional DH.

Both symmetric key and asymmetric key algorithms have unique advantages and disadvantages as identified in the following table:

	Symmetric key	**Asymmetric key**
Advantages	High performance/implemented in hardware	1. Does not require exchange of shared key 2. Keys required are linear with number of hosts

	Symmetric key	Asymmetric key
Disadvantages	1. Depends on exchange of shared keys 2. Need $n(n-1)/2$ key to communicate between hosts	1. Computationally expensive 2. Requires secure exchange of public key

Hash Functions Another key cryptographic primitive is the hash function, which takes some infinite length message, m, and produces a unique fixed length digest h, $H(m) = h$, as demonstrated in Figure 8.4. Cryptographic hash functions have some unique properties to ensure that they can be used. The examples include functions must be one way (i.e. that the attack cannot calculate M, even if they know m and $H()$). In addition to the one-way property, hashes should also be collision resistant, meaning an attack cannot find two message, $m1$ and $m2$, such that they both hash to the same digest, $(H(m1) = H(m2))$. These properties are particularly important to support a variety of authentication techniques discussed in later sections. The examples of secure hash functions include SHA-256, SHA-512, and SHA3.

Cryptographic Protocols Also, as identified in the previous sections, both symmetric and asymmetric cryptographic protocols have their own weaknesses. However, most cryptographic protocols will utilize a combination of multiple protocols to utilize the best attributes of both approaches. In this approach, asymmetric key algorithms, such as RSA or Diffie–Hellman, are used to exchange or agree upon a shared key, and then symmetric key algorithms are used to perform the resulting encryption. In this manner, individual hosts are not required to share large numbers of keys but can still utilize the improved performance of symmetric key algorithms.

IPSec is a common protocol to support VPNs, which provide a convenient way to communicate otherwise unsecure data over an untrusted network. IPSec operates at the IP layer and therefore encapsulates all the data. IPSec provides security by packaging all data into an encapsulated security payload (ESP), which is both encrypted and authenticated. IPSec sessions can be set up on a network-to-network approach, with VPN performing the encryption so the end nodes are not required to support this. However, before a security IPsec session

Figure 8.4 Hash functions.

can be established, encryption keys must be exchanged between client and sender. A Security Association (SA) is identified for each unique pairing of cryptographic keys and algorithms for session. The SA is then included in the ESP header to let the receiving device known how to perform the decryption. However, the parameters of the SA must first be established with a key exchange. The IPSec Key Exchange (IKE) and Internet Security Association and Key Management Protocol (ISAKMP) protocols are typically used to perform the key exchange and establish the SA.

8.4   Introduction to Encryption and Authentication

Authentication is used to maintain the integrity and authenticity of the data by preventing a malicious individual from spoofing messages or obtaining an interactive session with a remote resource. Because integrity is a more critical property for the power grid, authentication is important for most power grid communications, especially the SCADA telemetry and control messages.

A number of algorithms are used to provide authentication, including the symmetric and asymmetric key algorithms discussed in the encryption section but used in a different sequence. The symmetric key cryptography is used to enable message authentication codes (MACs), which are very efficient but required pre-shared keys. On the other hand, digital signatures, which are based on asymmetric key cryptography, therefore have worse performance.

8.4.1   Message Authentication Codes (MACs)

MACs utilize common symmetric key algorithms to provide integrity and authenticity instead of confidentiality. In this approach, the symmetric key algorithm is used in a slightly different manner, as we want to produce a unique value for that message, which is dependent on the key such that an attacker could not create a MAC that maps to a message of their choice. The overview of the MAC operation is provided in Figure 8.5.

Here, we can see that the function takes the key and messages and produces the MAC, which is then appended to the session. In this case, the encrypted message data are repeatedly fed back into the MAC algorithm until a digest based on the key and original message is produced. Because the digest is a function of the message and k, an attacker would have to crack the symmetric key algorithm before they could obtain the key, which is necessary if they want to produce a MAC value that corresponds to a message of their choice. Additionally, there are hashed MACs or HMACs that instead utilize hash functions instead of block ciphers to create the MAC with the key and message as inputs.

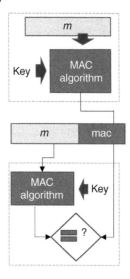

Figure 8.5 MAC algorithm.

8.4.2 Digital Signatures

While MAC utilizes symmetric key algorithms, digital signatures utilize public key cryptography to provide message integrity and authenticity. However, because of the computational expense of symmetric key algorithms, they cannot be used to sign an entire message. Therefore, digital signatures are traditionally assigned only the hash digest of the message.

There are multiple algorithms to perform digital signatures; probably, the most popular is RSA. Assuming a sender, s, has the public and private key pair (Prs, Pus) and a message m that they want to sign, the steps taken to produce the digital signature are as follows (in this section, we skip the mathematics of the RSA algorithm but provide links below to additional resources):

1. Create a hash, h, for the message, $H(m) = h$.
2. Sign the hash the sender's private key, $Sign(P_{r_s}, h) = d_{s_m}$.
3. Append the d_s to the message, $m|d_{s_m}$.

Once someone receives the message and digital signature, they can verify it by signing the signature with the sender's public key, $Verify(P_{u_s}, d_{s_m})$, and then verify that this equals the hash of the message. This works because the public and private keys are developed such that $Verify(P_{u_s}, Sign(P_{r_s}, h)) = h$, so that signing with the private key and verifying with the public key result in the original hash.

Figure 8.6 Certificate generation.

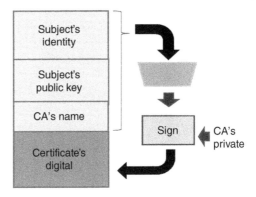

8.4.3 Certificates

While public key cryptography simplifies the need for shared keys, it still requires that individuals can distribute their public keys before they can communicate securely. Unfortunately, this distribution is vulnerable to tampering. Certificates depend on a trusted certificate authority (CA), which is a trusted third party to all communicating subjects. A certificate is a document that contains the following:

1. subject's identity
2. their public key
3. the CA's identity
4. a signed hash of the certificate by the CA's public key, as shown in Figure 8.6.

Instead of directly distributing a subject's public key, we now distribute the certificate, which contains the public key. Because the certificate is signed by the CA, as long as the receiving entity has the CA private key, it can verify that the public key in the certificate is actually the one that was presented to the CA. In order for an attacker to manipulate the certificate to add an incorrect key, they would have to create a signed hash that matches the malicious certificate; however, this requires breaking the digital signature function.

It should be clear that the certificate reduces the number of pre-exchanged certificates from the total number of communicating subjects to only one. Often the certificates can be deployed during the software development of deployment phase to address the challenge of distributing the certificate. Of course, this process moves a significant amount of trust into the CA, whose private key could not be used to create malicious certificates. A general overview of power system communication security requirements is provided in the following table:

Networks	Functions	Time reqs.	Security reqs.	Topology	Protocol
AMI					
Home area network (HAN)	1. Consumption monitoring 2. Consumer device/pricing interface	Minutes	Integrity – Medium Availability – Medium Confidentiality – High	Unicast	ZigBee
Field network	1. Report user consumption for billing 2. Update energy pricing 3. Meter maintenance/mgmt	Minutes	Integrity – High Availability – Medium Confidentiality –High	Multicast	ANSI C12.22, ZigBee
Distribution					
Distribution SCADA	1. Distribution automation 2. Fault detection, isolation, and service restoration 3. Support distributed energy resources (DER)	Seconds	Integrity – High Availability –High Confidentiality –Low	Unicast/ multicast	IEC 61850, DNP3, Modbus
Transmission/ generation					
Transmission SCADA	1. Communicate line voltage/ current data, operator controls 2. Support EMS functions	Seconds	Integrity – High Availability –High Confidentiality – Low	Unicast/ multicast	IEC 61850, DNP3, Modbus
Substation	1. Support protective relaying 2. Special protection scheme support	Milli– seconds	Integrity – High Availability – High Confidentiality – Low	Unicast/ multicast	IEC 61850
WAMS	1. Publish PMU readings 2. Obtain/process external PMU data	Milli– seconds	Integrity – High Availability – High Confidentiality – Medium	Unicast/ multicast	IEC 68150–90–5, C37–118
Inter–control center	1. Generation scheduling 2. Transmit grid status	Seconds	Integrity – High Availability – Medium Confidentiality – Low	Unicast	ICCP

8.5 Cryptography in Power Systems

While an earlier section provides introduction to traditional encryption technologies, there are many additional challenges implementing these technologies to protect power system communications. Specific examples of these challenges include the following:

1. *Performance and latency*: Many of the proposed technologies introduce unacceptable latency into the network communications. This is often a combination of older equipment with insufficient processing power or communication requirements that must occur in near real time.
2. *Legacy systems*: Many of the embedded devices used in substations do no support modern cryptographic functions.
3. *Long system lifespans*: The lifespan of a system within this domain is often over 10 years, as opposed to the 3- to 5-year span in most traditional IT environments; unfortunately, threats cannot be estimated out this far and it is likely that systems will be vulnerable to new threats as they arise.
4. *Large deployments*: Large utilities can have hundreds of substations, which require the deployment and management of a large number of keys and devices; therefore, techniques are required to reduce this overhead.

Modern power systems have a varying number of communication types that have varying requirements for communication latency. IEC 61850 identifies a variety of these types and their latency bounds as identified in the following table. This table shows that the fault isolation and protection functions have the lowest latency bounds of 3 and 10 ms. It is clear that these messages provide very little time to perform expensive encryption operations. Fortunately, many of these communications occur within a substation where encryption and authentication is less critical. In addition to the fault isolation and protection functions, certain routine automation functions may also have relatively low latency, e.g. below 20 ms. These messages may be problematic in scenarios where there is also noticeable propagation delay.

Functions	Message type	Delay (ms)
Fault isolation and protection	Type 1A/P1	3
	Type 1A/P2	10
Routine automation functions	Type 1B/P1	100
	Type 1B/P1	20
Measurement readings	Type 2	100
	Type 3	500

While this table explores the communication latency bounds of common power system messages, we have not yet explored the time to execute various cryptographic operations in the current hardware. The following table provides the result of recent studies exploring the time to execute many common cryptographic functions. This work was performed on a 2.8 GHz AMD processor and was evaluated on publisher/subscriber communications architecture. From the results, it is clear that both 128-bit AES and SHA 256 could be computed very quickly and are likely sufficient even for the power system application with the lowest latencies. However, the public key algorithms (RSA and DSA) had excessive delays, such that they may not be feasible for anything but the slow Type 2 and Type 3 messages, which do not require latency under 100 ms.

Algorithm	Pub (ms)	Sub (ms)	Total (ms)
128 bit AES	0.04	0.03	0.07
SHA-256	0.01	0.01	0.02
2048 bit RSA	59	2.04	61.04
1024 bit DSA	4.10	9.80	14.90

8.5.1 IEC 62351

The IEC 62351 standard specifies protocols for the encryption and authentication of power system communications. The standard identifies different standards based on the network protocol, and the following section explores the protocols specified by this standard.

Key exchange: The standard authentication approach is x509 certificate utilizing cryptographic key cryptography (e.g. RSA and DSA) to authenticate devices. In cases where trusted third parties are required, the protocols specify the use of the Group Domain of Interpretation (GDOI) key exchange protocol. The key exchange method is based on whether the protocol is based on the Transmission Control Protocol/Internet Protocol (TCP/IP) or Ethernet.

1. *TCP/IP*: The Transport Layer Security (TLS) protocol is recommended for the authentication and encryption of TCP/IP messages, along with the use of x509 certificates for both client and server [aaf]. While TLS provides strong security, it requires that a secure connection is first established between the sender and the receivers and therefore is not optimal in a scenario with low latency requirements.
2. *Ethernet*: The Generic Object Oriented Substation Events (GOOSE) protocol operates over Ethernet as it is not often routed outside of the substation. The standard suggested that RSA-based digital signatures using a SHA-256 message digest should be used for the communication. There is no requirement for encryption as these are unlikely to traverse an untrusted network.

8.5.2 DNP3 Secure Authentication (SA)

As discussed in an earlier chapter, DNP3 is one of the most popular protocols for SCADA communications. Unfortunately, this protocol was developed decades ago when cybersecurity risks were not major concerns; therefore, the protocol provides little security protections. To address this, a recent version (5) has introduced the concept of SAs, which extends the current protocol to provide authentication. DNP3 SA utilizes the challenge–response authentication protocol, which is different from the algorithms previously discussed in this section. The challenge–response requires the authentication of a random value (or *nonce*) to prevent an attacker from being able to replay old messages.

There is a significant difference in the importance of the various message types in SCADA. In DNP3, the message types, called Application Service Data Units (ASDUs), have a criticality value that is categorized as either Mandatory and Optional as identified in the following table. This categorization is used to determine whether the ASDU requires authentication. Here, we can see that the Mandatory ASDUs primarily include those functions that set some data points or perform operation of some control (e.g. opening a circuit breaker), while the optional ASDUs include the functions that monitor the various SCADA values.

Critical	ASDU	Related substation functions
Mandatory	Write, select, operate	Operate circuit breakers Change transformer tap position FACTS devices
Optional	Read, confirm	Current/voltage measurements Breaker status

There are two approaches DNP3 SA can use to authenticate ASDUs. The first is a challenge–response approach, which is traditionally used with message latency not a significant constraint. If latency is imperative, then an "aggressive" mode is available, which uses a simpler method. There are two approaches DNP3 SA can use to authenticate ASDUs, both of which assume that the sender and the receiver both share a symmetric key. The first is a challenge–response approach, which is traditionally used with message latency not a significant constraint. If latency is imperative, then an "aggressive" mode is available, which uses a more direct approach. To perform the actual authentication, DNP3 recommends either the SHA-HMAC or the AES-GMAC algorithm. Figure 8.7 displays two general approaches, which will be discussed as follows.

Before the message authentication can occur, first the DNP3 master and slave must establish a shared session key. In addition to the session key, they may also establish an update key that will be used to establish a new session key if they feel it is compromised. The protocol also allows the changing of the update key

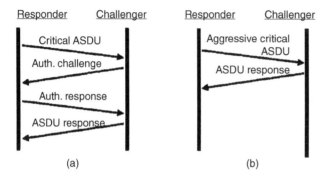

Figure 8.7 DNP3 authentication techniques. (a) Challenge-Response. (b) Aggressive.

through its symmetric or asymmetric key approach. In the symmetric key method, both must utilize a trusted third party and share a key. However, in the asymmetric approach, each device will utilize a public/private key pair to perform the key exchange.

Challenge–response mode: The challenge–response mode is demonstrated in Figure 8.7. Here, we can see that it takes two round-trip times to send the ASDU. The first message is sent by the DNP master (responder) to the slave (challenger) and will contain the critical ASDU. To ensure that this is a valid message, a challenge–response will then be initiated by the challenges to ensure about the authenticity of the ASDU. The challenge message will include a variety of fields including random string, sequence number, and algorithm (e.g. SHA-HMAC and AES-GMAC). When the responder receives this, they will utilize the shared key, algorithm, and random number to compute a response-based unique to that message and then deliver it back to the challenge. When receiving the response, the challenger can use the MAC value as it also knows the same set of values (e.g. algorithm, shared key, and random number). The challenger will then send the appropriate ASDU response based on whether the ASDU was validated.

Aggressive: Many communications with low latency restrictions may not have time to perform two round-trip communications for each session. The aggressive mode addresses this by enabling authentication within one Round Trip Time as demonstrated in Figure 8.7. Here, the challenger sends an aggressive ASDU packet to the challenger. This aggressive ASDU includes the MAC based on the ASDU fields and then sends the MAC with the aggressive ASDU. The challenger will then verify the MAC and send the appropriate response. While this approach is faster, it does not involve the transmission of a random number, and therefore, the aggressive ASDU may be more vulnerable to a replay attack than challenge–response methods.

8.6   Access Control

While authentication focuses on verifying the sender of the message and verifying its integrity, there still remains a challenge regarding what messages the devices will accept and what functions the various system users should be able to access. Access control provides a mechanism to limit the data and functions that can be accessed by the various users, devices, and programs. Because many smart grid devices are accessed by many different types of users (e.g. engineers and administrators), and devices, it is important that each entity is only restricted to the minimal set of permissions required for them to execute their tasks. Access control is traditionally described as a set of subjects, such as users and programs, accessing a set of objects, which could be files, processes, or computer resources (e.g. network sockets). An overview of access control is shown in Figure 8.8.

There are various access control mechanisms that can be implemented to monitor and protect access to system resources. Discretionary access control (DAC) is the most popular method and is the default on most operating systems (e.g. Windows and Linux). In this method, the owner of any file has the ability to determine who can read, write, or execute it. However, this approach has many limitations as unique privileges must be available every time there is a new employee, job responsibility, or they are terminated, which is an error-prone and time-intensive process. To address these, the concept of role-based access control (RBAC) was introduced.

RBAC introduces the concept of roles based on specific job functions that would be common to an organization, such as a utility. Each role is then assigned to a set of rights or privileges that specify the set of objects they are able to read, write, or execute. Finally, each user is assigned to one or more roles based on their various job responsibilities. This creation of a role abstracts the individual assigning of privileges to users and simplifies the assignment of privileges.

8.6.1   RBAC in IEC 62351

To alleviate privilege assignment in the power grid, IEC 62351 introduces a standard RBAC model based on common roles and objects that would be available. The following table provides an overview of the roles and privileges model for IEC

Figure 8.8   Access control overview.

62351. The privileges include the ability to view, read, control, reporting, and data set. There are three roles: viewer, operator, and engineer.

	Privilege				
Roles	View	Read	Control	Reporting	Dataset
Viewer	X		X		
Operator	X	X	X		
Engineer		X		X	X

Here, we can see that the viewer only has access to the view role. The operator has additional access to read operational system data and then can also send control commands to various system components. Finally, the engineer's role is not allowed to submit any control functions but should be able to explore reporting and data set information to understand system settings. Figure 8.9 provides an example access control scheme with two users, UserA and UserB, where UserA is only assigned to the viewer's role and UserB is assigned to both the operator and engineer's role and can switch between them in different sessions.

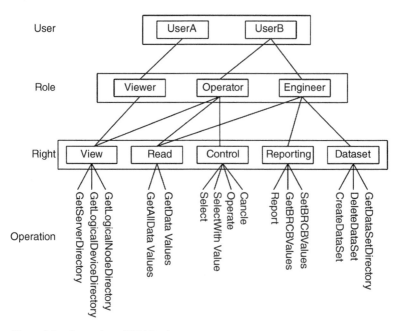

Figure 8.9 Examples of RBAC roles.

8.7 Summary

This chapter identified some of the key terms in cybersecurity, including how threats, risks, and vulnerabilities must be incorporated into understanding the need to protect power system cyber assets.

The chapter also introduced a number of key security mechanisms necessary to protect substations, including encryption, access control, and authentication. While these are technology, there are various standards, such as IEC 62351, that discuss their usage in power systems.

8.8 Problems

1 An attacker installs malware that manipulates sensitive database records, what property is degraded?
   A Confidentiality
   B Integrity
   C Availability
   D Authenticity

2 Which encryption cipher provides the greatest security assuming that it is implemented correctly?
   A AES
   B One-time pad
   C RSA
   D MD5

3 Which of the following is not a common symmetric key algorithm?
   A AES
   B RSA
   C DES
   D 3DES

4 What is a primary weakness of modern symmetric key ciphers?
   A Cost
   B Security
   C Key distribution
   D Performance

5 What is a key difference between the IPsec and TLS security protocols?
A Payload size
B Symmetric key algorithms
C TCP/IP layer
D Hash algorithms

8.9 Questions

(1) This chapter discussed the need for confidentiality, integrity, and availability in a SCADA system. Explain the importance of these three properties (high, medium, and low) in a modern smart meter.

(2) State and explain both Kirchhoff's laws and Kerckhoffs's laws and their similarities and differences.

(3) Explain why GOOSE messages are not commonly encrypted; is this a significant vulnerability?

(4) Explain the advantages and disadvantages of symmetric and asymmetric key cryptography. How do these apply to common power system communications?

(5) For the following types of communication: (a) protection, (b) automation functions, and (c) measurement readings, identify what authentication scheme would be most applicable and why.

Further Reading

Anderson, R. (2020). *Security Engineering: A Guide to Building Dependable Distributed Systems*. Wiley.

Barker, E., Barker, W., Burr, W. et al. (2007). NIST special publication 800-30. *NIST Special Publication 800-30*.

Geer, D.E. Jr. (2008). Complexity is the enemy. *IEEE Security and Privacy* 6 (6): 88–88.

Hahn, A. and Govindarasu, M. (2011). An evaluation of cybersecurity assessment tools on a SCADA environment. *2011 IEEE Power and Energy Society General Meeting*, pp. 1–6. https://doi.org/10.1109/PES.2011.6039845.

Mulligan, D.K. and Schneider, F.B. (2011). Doctrine for cybersecurity. *Daedalus* 140 (4): 70–92.

Savage, S. and Schneider, F.B. (2009). Security is not a commodity: the road forward for cybersecurity research. Retrieved May 31: 2010.

Stouffer, K., Falco, J., and Scarfone, K. (2011). Guide to industrial control systems (ICS) security. *NIST Special Publication 800-82*, p. 16.

9

Network Attacks and Protection

This chapter will introduce the broad types of network-based attacks, especially the ones typically seen in the power systems domain. These types of attacks have various implementation techniques, and we will study the concept behind these attacks and some typical implementation techniques. A variety of security methods can be used to detect and mitigate these attacks. These methods and their implementation techniques will be discussed.

9.1   Attacks to Network Communications

In this section, we will discuss the concepts of the two most common attacks on network communications – denial-of-service (DoS) and spoofing. We will discuss in detail the concept behind the attacks, some implementation techniques, and examples.

9.1.1   Denial-of-Service (DoS) Attack

One of the most common network attacks is the DoS attack, typically called the DoS attack. A DoS is defined as an action that prevents or impairs the authorized use of network systems or applications by exhausting resources such as the central processing unit (CPU), memory, bandwidth, and disk space. Some techniques that are used to implement the DoS attack are discussed as follows:

1. *Malformed packet*: A malformed packet can trigger some software vulnerability or weakness that causes a system crash and denies the authorized user the use of specific resources needed.
2. *Flooding*: Flooding refers to the overwhelming of system resources such as network bandwidth or CPU speed.
3. *Protocol-based attacks*: In this type of attack, the attacker would manipulate the protocol state – a typical example of TCP reset to perform a DoS attack.

Cyber Infrastructure for the Smart Electric Grid, First Edition.
Anurag K. Srivastava, Venkatesh Venkataramanan, and Carl Hauser.
© 2023 John Wiley & Sons Ltd. Published 2023 by John Wiley & Sons Ltd.

Other implementations of the DoS attack can be

1. a distributed denial-of-service attack called as DDoS,
2. a reflection/amplification attack,
3. a slashdot/flashcrowd type of attack that is non-malicious and occurs because the network state is being overwhelmed by unusual and non-typical network activity.

9.1.1.1   Flooding

The goal of the flooding attack is to overload the capacity of the network, in the case of a targeted host, to overwhelm the capabilities of the system of interest. In the case of a network, the flooding attack will aim to consume the physical resources enabling the network, such as the bandwidth. In the case of a host system, the attacker might overwhelm the ability of the system to process data by sending huge amounts of data. There are many techniques or types of flooding attacks such as follows:

1. *ICMP flood*: A flooding attack using ICMP might involve the attacker sending a large number of ICMP "Echo Request" messages. This type of attack might not be effective for current networks as they are often filtered. A variation of this attack is flooding using "ICMP Destination Unreachable" messages, as they are not commonly filtered by mitigation systems.
2. *UDP flood*: This technique involves sending l arge UDP packets to system administration functions such as DNS (domain name system). The UDP protocol is connectionless; hence, the attacker has the flexibility of launching an attack as soon as getting access to the network, unlike TCP protocol. TCP presents a handshake overhead, which makes UDP an attractive proposition for the attacker.
3. HTTP flood: This technique involves attacking more recent mission-critical applications that are deployed as web services. This technique involves sending legitimate HTTP GET/POST messages to the web server, which might overwhelm the server.
4. TCP Syn flood: This is a well-established attack method that is more of threat on legacy and unpatched systems. In this technique, the attacker sends a large number of TCP "SYN" packets to the server. These packets are used to establish a TCP connection. The server creates a half-open TCP connection state and sends a Syn-ACK. If the client does not send back an "ACK" message, the server continues to allocate memory for a connection that is not active. The attack exhausts the finite number of half-open connections that is feasible for a given system. This technique is mitigated in newer implementations of the TCP protocol by removing the SYN message from the SYN queue after

sending the Syn-ACK and stores the SYN message as a "SYN Cooking," which encodes information such as IP addresses, ports, and sequence numbers.

9.1.1.2 Malformed Packet

A malformed packet-based DoS can be the result of malicious activity by an attacker or due to faulty code. The malformed packet is also referred to as a "crafted packet," when it is a software exploit created by an attacker. The malformed packet aims to cause the software or an operating system crash. This is usually a result of a software vulnerability or error. Some examples of this attack technique include the "Ping of Death," and the "Teardrop" attacks. In the Ping of death, an ICMP ping packet with size greater than 2^{16} bytes is crafted, which violates the protocol. This caused buffer overflow/crashing of older Windows/Unix systems. In the Teadrop attack, the process of reassembly of fragmented IP packets is targeted. The attacker would craft overlapping fragments of the IP packets, which would cause the operating system to crash.

9.1.1.3 Reflection

In the reflection attack, the attacker spoofs the IP address of an intermediate system. The intermediate system responds to the messages from a target system, usually a server. This causes the server or the target system to drop connections/create unwanted responses to a victim. This causes the victim to misidentify the attacker. The concept of reflection attacks can be demonstrated using a TCP handshake event. In the normal case of Figure 9.1, consider a client initiating a TCP handshake by sending a SYN packet. The server receives the SYN and

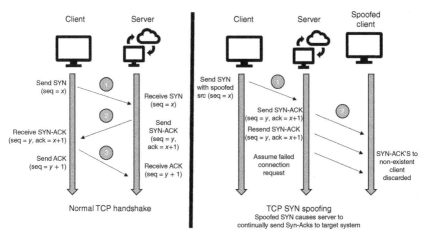

Figure 9.1 Reflection attack.

responds back with a SYN-ACK and the connection is created. In the case of a reflection attack, the attacker initiates a SYN with spoofed credentials. The server then responds with the Syn-ACK to the victim ("spoofed client"), which then discards the SYN-ACK. The server keeps resending the Syn-ACK to the victim host, which might cause it to malfunction.

An attacker might choose to use a reflection attack to avoid detection, as the server is creating unwanted additional traffic. In addition, attack bandwidth can be amplified by using a single attacker machine to overwhelm the server and multiple victims. This type of attack is well suited for TCP handshake and certain UDP functions such as DNS requests, SNMP, or NTP messages. In the case of UDP, the attacker does not have to set up sessions and might have easier access to the network.

9.1.1.4 DDoS

DDoS stands for distributed denial-of-service. The DDoS is among the most popular cyber attack techniques, along with malware and ransomware. The DDoS technique utilizes a large number of attacking systems, which increases the amount of traffic sent by the attack. This sort of attack is more difficult to mitigate than a typical DoS attack as it is harder to differentiate between normal and attack traffic. The control for a DDoS attack can be centralized or distributed. The centralized implementations typically use "botnets" – which are a large group of network-connected devices operated by a botnet operator. Centralized DDoS can be easier to detect and mitigate, as they are controlled only by a single entity. In the case of distributed DDoS, the attacks are launched by multiple entities, who can be a group of individual attackers (such as the hacker group Anonymous) or nation states. These entities also use botnets, but their control is distributed, making it harder to detect and mitigate the source of the attacks. DDoS attacks often use a combination of techniques to amplify the amount of data being sent to a target. A common tactic is to use a reflection/amplification attack on a DNS server, which would in turn generate much more traffic than is capable by just the botnet alone.

9.1.2 Spoofing

Historically, spoofing relied on the fact that IP (Internet Protocol) was developed without an authentication mechanism, in the 1970s. The source address of a message can be spoofed so that the receiver thinks that the sender was someone else. This concept is still seen frequently in attack implementations such as DoS. Multiple security mechanisms have been developed to mitigate IP spoofing. This includes newer routers, which may filter packets with incorrect source IP addresses. Another common mitigation strategy is the use of IPsec, which is an improved version of the standard IP, and provides authentication of IP packets. The upcoming standard IPv6 also provides default support for IPsec. Some other typical spoofing techniques are discussed as follows:

9.1.2.1 ARP Spoofing

ARP that stands for Address Resolution Protocol is a critical part of ensuring that the packet from a source reaches the correct destination. In the TCP/IP stack, a packet is directed to its destination by its IP address. The ARP is used to resolve the IP address by connecting them to the MAC address (link layer) of a particular device. While IP address can be reused under different subnets, the MAC addresses are unique and the ARP ensures that the packet being sent is routed to its proper destination. The problem with the ARP, which is similar to IP, is that it does not have authentication by default. Hence, an attacker can create a malicious ARP response by claiming to be the system with the IP requested by the sender. In this case, the ARP resolution turns out to be a race between the actual target with the proper IP and MAC address and a malicious actor who might be spoofing a particular IP. In power grid applications, this can be mitigated by creating static pre-determined ARP tables on the hosts and network switches and only allowing authenticated changes. The ARP spoofing attack is demonstrated in Figure 9.2.

9.1.2.2 Other Spoofing

In addition to IP and ARP spoofing, other types of network packets can also be spoofed. For example, TCP is a stateful connection, which means that it tracks

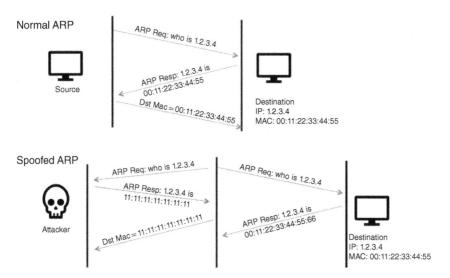

Figure 9.2 ARP spoofing attack.

packets by using sequence numbers and acknowledgment numbers, and packets with incorrect sequence numbers will be rejected. The sequence number is a 2^{32} number. Attackers can take advantage of this by initiating a TCP connection with the standard starting sequence number. This can be mitigated by creating randomized initial sequence numbers (ISNs) to prevent the user from guessing the initial number. However, the attacker can still get around this by eavesdropping a TCP packet, learning the sequence number and then responding faster than the destination with a spoofed packet.

The Border Gateway Protocol (BGP) manages how packets get routed from network to network on the Internet through the exchange of routing and reachability information among routers. BGP packets can also be spoofed by attackers to advertise false routing paths to hijack traffic and redirect users to malicious websites. This has been performed several times by various nation states to hijack traffic from popular websites such as YouTube. It has also occurred non-maliciously in the case of Facebook because of a misconfiguration in the BGP packet.

A user can be misdirected to malicious websites also by DNS spoofing. While BGP provides the most efficient route to reach from source to destination on a network, DNS provides the location of the destination that is used by the BGP to determine its route. Hence, BGP spoofing can be used to misdirect traffic through malicious servers, while DNS spoofing can enable the attacker to receive privileged information. The original DNS protocol did not have any authentication, and attackers could spoof DNS responses to get the user to visit a different system. This could take the form of either a man-in-the-middle (MITM) attack to simply manipulate the DNS response or it could only be a spoofing only attack where the attacker has no knowledge of the current traffic. DNS spoofing is mitigated in current implementations using Domain Name System Security Extensions, which prevent attackers from manipulating or poisoning the responses to DNS requests. DNS query IDs are also randomized to avoid potential compromises.

9.2 Mitigation Mechanisms Against Network Attacks

Mitigation mechanisms can be broadly grouped under three categories:

1. Cryptographic protocols and solutions
2. Firewalls and logical separations
3. Intrusion detection mechanisms

We will discuss on all of these mechanisms in detail.

9.2.1 Network Protection Through Security Protocols

In current implementations of the TCP/IP stack, security enhancements have been deployed across the various layers to enhance its security features, as shown in Figure 9.3.

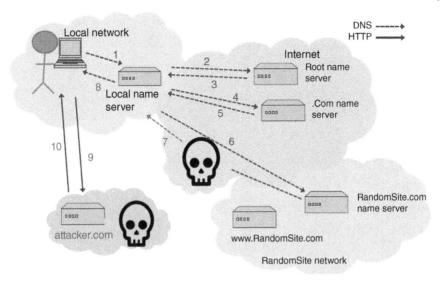

Figure 9.3 Security enhancements in the TCP/IP stack.

These security enhancements are necessary to securely communicate over an untrusted network and enforce security practices even on trusted or protected networks. They provide enhanced confidentiality, integrity, and authentication in combination across the layers as compared to previous deployments. These enhancements are largely driven by the cryptographic mechanisms discussed in Chapter 8.

9.2.1.1 TLS

The original implementation of TLS was called Secure Socket Layer (SSL). It was originally designed to support secure HTTP protocol, HTTPS. There also exists a UDP equivalent called Datagram TLS, which provides the same service for UDP-based communications. Currently, TLS is used to secure many other application layer protocols. TLS mainly provides authentication for messages by using MAC addresses and ensures confidentiality of messages by using encryption schemes. TLS generally uses asymmetric encryption techniques, although this can be configured as required by the deployment. The current version of TLS succeeded SSL 3.0, which had a weak key generation mechanism in its encryption. The current version is government approved based on NIST SP 800-52 and uses SHA-2 (256 or 512 bit) hash functions. The TLS protocol stack is shown in Figure 9.4.

TLs use a stateful connection with a handshake that is used to agree upon several parameters in the state, including (i) TLS version, (ii) ciphers, (iii) certificates, (iv) pre-master secret, (v) master secret, and (vi) session ID. A typical TLS handshake is shown in Figure 9.5.

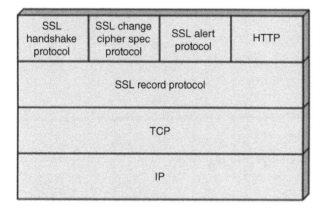

Figure 9.4 TLS protocol stack.

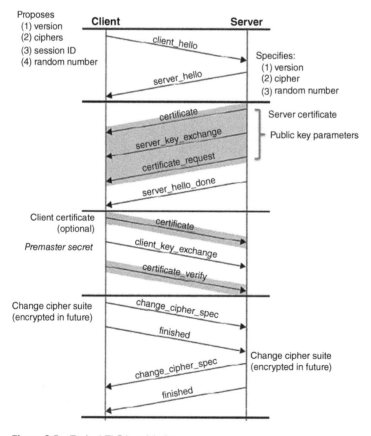

Figure 9.5 Typical TLS handshake.

TLS ciphers contain a set of cryptographic algorithms necessary to perform the following functions:

1. Key exchange algorithm, such as RSA or Diffe–Hellman
2. Bulk encryption algorithms, such as Stream (RC4) and Block (3DES, DES, AES, etc.)
3. Data authentication algorithms, such as MAC algorithms and HMAC with (MD5, SHA1, and SHA256)

9.2.1.2 IPsec

IPsec operates primarily on the Internet Protocol layer and is commonly used to build virtual private networks (VPNs). These VPNs can be host to host, network to network, or host to network. IPsec provides encryption and authentication at the network layer and has several functions:

1. Security associations (SAs), including algorithms and parameters used in encryption
2. Authentication Header (AH), which provides connectionless authentication and integrity
3. Encapsulated Security Payload (ESP), which provides confidentiality, authentication, and integrity

The advantage of IPsec is that it provides confidentiality, integrity, and authentication guarantees for all the IP packets, which is all routable traffic through a host. As it is based on open encryption technologies, it is open to users and moves the cryptographic processing to the network routers/devices rather than the individual host. IPsec deployments rely on the use of SAs. A security association is the establishment of shared security attributes between two entities using IPsec to support secure communication. A SA may include attributes such as cryptographic algorithms and encryption key to be passed over the connection to secure the communication. SAs are one way, and therefore, two SAs need to be created for two-way communication. SA establishment is done using the Internet Security Association and Key Management Protocol (ISAKMP). ISAKMP defines procedures for authenticating a communicating entity, creation and management of SA, key generation mechanisms, and threat mitigation for IP layer attacks such as denial-of-service or replay attacks. The IPsec ESP can operate in two modes depending on whether the communication is one way (transport mode) or two way (tunnel mode). These modes are illustrated in Figure 9.6.

How Does TLS and IPsec Differ? TLS and IPsec utilize differing mechanisms for connection establishment and trust establishment. IPsec uses pre-established SAs to agree on key conditions such as ciphers, encryption keys, and cryptographic algorithms. On the other hand, TLS utilizes its handshake to negotiate these

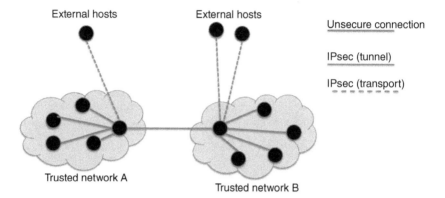

Figure 9.6 IPsec ESP modes of operation.

terms between the client and server to arrive at a consensus. There are pros and cons to both approaches – having the pre-established terms increase overhead before connection, while agreeing terms during handshake leaves it vulnerable to MITM and malicious security "downgrade" attacks. For trust establishment, IP uses the pre-established terms from the SA, while TLS utilizes trusted authorities such as trusted certificate authorities (CAs) or pre-shared certificates. *IEC 62351* For the smart grid, the IEC 62351 standard provides data and communication security standards. It provides specifications for key management in IEC 62351-9, which mandates X509 certificates for devices, and Group Domain of Interpretation (GDOI), which uses symmetric key management and trusted key servers for authentication. The standard also uses TLS for message encryption and RSA-based digital signatures for message authentication.

9.3 Network Protection Through Firewalls

A firewall is a network security construct that monitors and filters the incoming and outgoing network traffic based on a pre-established security policy. A firewall is essentially the barrier that sits between two sections of a network. A general firewall is shown in Figure 9.7.

Firewalls are primarily used to separate more critical and less critical networks and also restrict visibility to Internet on some networks. It is also used to enforce the desired security policies on traffic flows on the network. It does this by creating a single system through which all traffic on the network *must* pass through. The traffic data coming into the flows are divided into two types:

1. *Ingress*: data coming into the network
2. *Egress*: data leaving the network

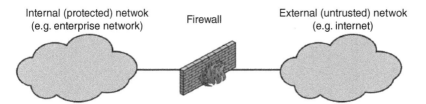

Internal (protected) netwok
(e.g. enterprise network)   Firewall   External (untrusted) netwok
(e.g. internet)

Figure 9.7 Firewall showing separation between the critical enterprise network and an untrusted network.

Table 9.1 Packet filtering rules – an example.

Rule	Direction	Src addr	Src port	Dst addr	Dst port	Prot	Conn state	Action
1	In	External	—	1.2.3.4	20,000	TCP	New, established	Permit
2	Out	1.2.3.4	20,000	External	—	TCP	Established	Permit
3	Both	Any	Any	Any	Any	Any	New, established	Deny

A firewall can be deployed either as a packet filtering mechanism, which works on stateful inspection or packet filtering, or as an application layer proxy. Both deployments work on different layers in the TCP/IP stack.

Packet filtering generally operates at the network, Internet, and transport layers. Packet filtering can be setup for various configuration such as a **"Default policy"**, where if a packet does not match a rule, it would be discarded or dropped, and forwarded or allowed otherwise. The rules that match the received packets can include parameters such as (i) source/destination IP, (ii) source/destination port, and (iii) protocol. If the packet filtering is aware of the state of TCP connections, it is called as "stateful inspections." An example of packet filtering rules is shown in Table 9.1.

In the example shown, the firewall only allows traffic to the DNP slave (IP:1.2.3.4, TCP port 20,000). External denotes the IP range for external systems.

In the case of an application layer firewall, it works primarily on the application layer protocol, such as SCADA applications or Web applications. In this case, it be set up to filter only email traffic by only allowing traffic to SMTP ports or be setup to filter DNP3 objects or IEC 61850 GOOSE messages to assign them with higher priority.

9.4 Intrusion Detection

An intrusion detection system (IDS) is a system that detects unauthorized access to networks and systems and produces alerts when detected. IDSs usually have

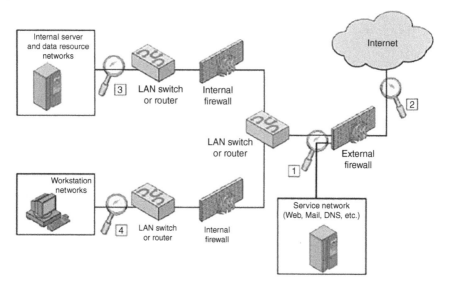

Figure 9.8 Intrusion detection systems.

sensors and an analyzer. The sensors are used to collect information from the system, which could be network packets, system log files, system calls, etc. The analyzer is an algorithm or a detection mechanism that receives inputs from the sensors and analyzes it and creates alerts. A general model of an IDS is shown in Figure 9.8.

IDSs are usually classified into these major types:

1. *Host-based intrusion detection system (HIDS)*: sensors collect data from hosts for malicious processes, network stack activity, modified files, etc.
2. *Network-based intrusion detection system (NIDS)*: sensors collect data from network
3. *Hybrid*: combine information from both network and host sensors

Another component of the IDS is the analysis that goes into detecting malicious behavior and creating alerts. This analysis could be performed in a number of different ways as follows:

1. *Anomaly-based*: compare current data to collection of past data, assumes deviation from past patterns (or anomalies) are attacks
2. *Signature-based*: use a set of known attack patterns that are compared with the current sensor data (e.g. Snort)
3. *Specification-based*: create "specifications" of known, correct system operation

9.4.1 Anomaly-Based Detection

The underlying concept of anomaly detection is to develop a model for normal behavior and compare that with the incoming events. If the incoming behavior differs from the normal expected behavior above a certain threshold, an alert is generated. The normal operational behavior can be modeled through several mechanisms, such as creating statistical model behaviors or using machine learning models to learn expected behavior from historical data. The advantage of such an approach is that it can detect new or unknown attack vectors, as they would still fall outside the expected behaviors. However, this can also result in many benign anomalies – such as rare operating phenomena such as network reconfiguration, system upgrades, new programs, etc. This can lead to the problem of excessive false positives, which is also called base rate fallacy. A false positive is represented in Figure 9.9.

A good IDS requires small false positives, which ensures that money/resources are not wasted investigating a non-attack. It also requires small false negatives, so that it does not miss attacks that might result in violations to the security policy. The overall base rate fallacy needs to be small to ensure that the IDS detects a small number of intrusions as compared to a large number of regular non-malicious traffic. These characteristics can be represented using a receiver operating characteristics (ROC) curve, which plots the true positives (Y-axis) against the false positives (X-axis).

9.4.2 Signature-Based Detection

Compared to the model-based approach of the anomaly-based detection, signature-based detection works by maintaining a collection of known patterns of malicious data and compares the incoming network traffic to these known malicious patterns. The obvious weakness of this approach is that it can only detect those patterns that have already been pre-defined, i.e. it can miss novel zero-day type attacks. On the other hand, it usually has a low false-positive rate, especially

Figure 9.9 Attack detection cases.

	Attack identified	
	Yes	No
Attack present — Yes	True positive	False negative
	No	False positive	True negative

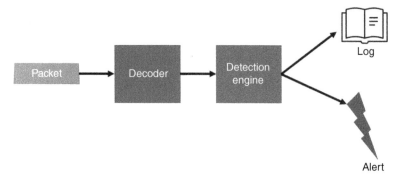

Figure 9.10 Snort IDS.

if the rules are created properly. The known attack patterns can again be defined by system operators using modeling approaches such as statistical modeling or by observing and identifying attack patterns from historical databases. An example of this type of IDS is the popular Snort IDS.

Snort is an open-source, signature-based IDS, which is usually deployed at the network level (NIDS). It can operate in passive or inline modes, where it will either only detect attacks or also actively block packets that match attack patterns. It has three components – (i) decoder, which decodes protocol layers and packet structures for analysis, (ii) detection engine, which analyzes the packet against a set of rules or patterns, and (iii) logger/alerter, which performs the appropriate action as necessary. A representative block diagram is shown in Figure 9.10.

Snort rules can be set up to identify suspicious patterns based on user parameters such as protocol, port, IP address, or direction of traffic flow. For example, the Snort rule could be

```
tcp any any $\rightarrow$ 192.168.1.0/24 111
```

Once a packet has been flagged by the IDS, various actions can be performed such as *alert, log, pass, drop, reject, activateaction*, etc. Snort also has several options that can be deployed to monitor packet payload information such as content, *pcre, header*, etc., or can look at non-payload information such as *ttl, seq,* and *ack* messages.

9.5 Summary

In this chapter, we discussed the various network attacks and their mitigation mechanisms. We studied the implementation of several network-based attacks such as spoofing, DoS, and other types of attacks. We also studied implementation-based mitigation schemes such as TLS and IPsec. In addition, network protection based on firewalls and logical separations and IDSs were studied.

9.6 Problems

1 Tripwire is a software tool intended to assure the integrity of system files by detecting unexpected modifications (such modifications are often a sign of rootkit activity). One version of Tripwire reads the names of the directories to be protected from a configuration file. For each file in the specified directories, Tripwire computes its hash value and stores it in a database. What property must this hash function have?
A pre-image resistance
B second-pre-image resistance
C collision resistance
D all of the above

2 SYN cookies (http://cr.yp.to/syncookies.html) are used to mitigate SYN flooding attacks. Which of the following is not used in computing a SYN cookie?
A Timestamp
B TCP flags
C Maximum segment size
D Port number

3 Suppose you have an intrusion detection system detecting a computer virus on the network with 90% accuracy. Precisely, the IDS detects a connection transferring a virus as an attack with 90% probability and a benign connection as an attack with 10% probability. When 1% of the connections contains a virus, what is the probability that a connection flagged by the IDS as an attack is actually benign?
A 0–20%
B 21–50%
C 51–90%
D 91–100%

4 What is not a common IDS analysis approach?
Anomaly based
A Specification based
B Protocol based
C Signature based

5 How should you categorize an event where an intrusion detection system raises an alert in response to an actual attack against a system?
   A True positive
   B True negative
   C False positive
   D False negative

9.7 Questions

(1) What is a flooding attack? Describe its variations.
(2) What is a spoofing attack? Describe how ARP spoofing is carried out.
(3) Explain network protection through TLS and IPSec and how they differ.
(4) Explain network protection through firewalls and draft firewall rules for (i) computers in the enterprise network in a substation and (ii) devices on the OT network, such as a protection relay.
(5) Explain the two methods of intrusion detection and how they are deployed.

Further Reading

Hahn, A. and Govindarasu, M. (2011). Cyber attack exposure evaluation framework for the smart grid. *IEEE Transactions on Smart Grid* 2 (4): 835–843.

Humayed, A., Lin, J., Li, F., and Luo, B. (2017). Cyber-physical systems security—a survey. *IEEE Internet of Things Journal* 4 (6): 1802–1831. https://doi.org/10.1109/JIOT.2017.2703172.

Li, X., Liang, X., Rongxing, L. et al. (2012). Securing smart grid: cyber attacks, countermeasures, and challenges. *IEEE Communications Magazine* 50 (8): 38–45.

Peng, C., Sun, H., Yang, M., and Wang, Y.-L. (2019). A survey on security communication and control for smart grids under malicious cyber attacks. *IEEE Transactions on Systems, Man, and Cybernetics: Systems* 49 (8): 1554–1569. https://doi.org/10.1109/TSMC.2018.2884952.

Stellios, I., Kotzanikolaou, P., Psarakis, M. et al. (2018). A survey of IoT-enabled cyberattacks: assessing attack paths to critical infrastructures and services. *IEEE Communication Surveys and Tutorials* 20 (4): 3453–3495. https://doi.org/10.1109/COMST.2018.2855563.

Zhu, B., Joseph, A., and Sastry, S. (2011). A taxonomy of cyber attacks on SCADA systems. *2011 International Conference on Internet of Things and 4th International Conference on Cyber, Physical and Social Computing*, pp. 380–388. IEEE.

10

Vulnerabilities and Risk Management

In this chapter, we will discuss system vulnerabilities that come in various categories and some security mechanisms that are used to mitigate them. These security mechanisms are driven by security evaluation methods, which play a key role in understanding the threat from a particular vulnerability. Various mechanisms for security evaluation will be discussed, and related governmental and regulatory efforts toward creating a framework for evaluating system threats and risks will be studied.

10.1   System Vulnerabilities

System vulnerabilities can occur due to various reasons, and these vulnerabilities can be exploited by malicious actors, leading to threats and risks to the system. Vulnerabilities are described by various characteristics such as the time they are introduced to the system, system components from which they originate or affect, and type of impact on the system.

Based on the time they are introduced, vulnerabilities could be due to the following:

1. *Design*: System design might be incorrect or there might be insufficient system requirements
2. *Implementation*: Errors in code or improper detail in implementation
3. *Operational*: Software might not be suitable for the environment on which it is run

Vulnerabilities can also be classified by the system components on which they are manifested as (i) software vulnerabilities, (ii) hardware vulnerabilities, and (iii) network vulnerabilities. The effect of such vulnerabilities can be studied under the confidentiality, integrity, and availability (CIA) framework. We will study these classifications in detail.

Cyber Infrastructure for the Smart Electric Grid, First Edition.
Anurag K. Srivastava, Venkatesh Venkataramanan, and Carl Hauser.
© 2023 John Wiley & Sons Ltd. Published 2023 by John Wiley & Sons Ltd.

10.1.1 Software Vulnerabilities

Software vulnerabilities are bugs present in the code that could have various etiological reasons such as programmer error (including typos), design errors, or improper deployment. Figure 10.1 shows the total number of lines of code (LoC) in a modern SCADA/EMS system, which is approximately 2,000,000 for the operating system (OS) and 60,000,000 for the actual power system application. Of these, there are usually 2–3 "bugs" per 1000 LoC, as estimated by various organizations such as NIST. These conclusions are derived from software quality testing studies across a wide variety of software from different domains, which is a conservative estimate. It is important to understand the distinction between bugs and vulnerabilities in the security context. A bug is an instance of a system that does not perform as expected, while a vulnerability is a bug that allows a malicious actor to exploit to create failures. Of the 2–3 bugs per 1000 LoC, about 5% develop into software vulnerabilities that work out to approximately 0.1 vulnerabilities per 1000 LoC. Typically, these vulnerabilities are discovered during testing phases or by security experts performing procedures such as fuzz testing. Not all vulnerabilities that are discovered are "patched" or fixed, as the patching process is often expensive and involves pushing updates to remote machines. A small fraction of vulnerabilities are left unpatched and are documented so that the user is aware and can take preventative actions as deemed necessary. However, a large portion of the

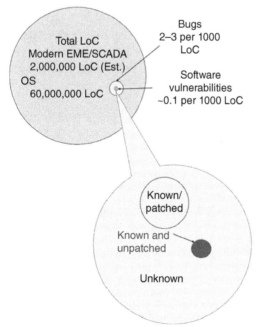

Figure 10.1 Bugs and vulnerabilities in software.

vulnerabilities remains undiscovered and provide opportunities for attackers to launch "zero-day" attacks, which are software exploits not known to the consumer and vendor.

The SANS Institute is a private U.S. for-profit company founded in 1989 that specializes in information security. Working with MITRE, which is another company focused on cybersecurity, SANS releases a list of top 25 most dangerous software errors, and the list is updated periodically. Some of the most relevant threats to the cybersecurity of the smart grid include the following:

1. Use of hard-coded credentials
2. Improper authorization
3. Improper authentication for critical functions
4. Improper encryption of sensitive data
5. Buffer copy without checking the size of input ("Classic buffer overflow")
6. Improper Restriction of Operations within the Bounds of a Memory Buffer
7. Improper Control of Generation of Code ("Code Injection")

Buffer overflow is a particularly interesting vulnerability, as it is a very specific type of exploit that occurs because of improper design. This is an exploit that happens mainly in controllers or devices that receive input or either from a sensor or from the user. The general structure of a run-time memory showing the subdivisions of stack, heap, static memory, code, and free memory is shown in Figure 10.2.

When the user or a device sends an input to the stack, the packet needs to be received and processed such that it only occupies a particular position in the stack or the free memory. If the input bounds are not specified, a malicious actor can create a crafted input such that they get access to unauthorized memory locations. If the attacker is allowed to take up additional space in the buffer without restrictions, they can send a large input bigger than typical buffer sizes that enables them to manipulate the **ret** value, which is the value returned by the process back to the user. This is illustrated in Figure 10.3.

Figure 10.2 Run-time memory of a process.

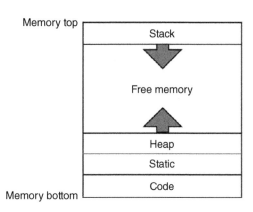

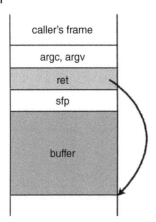

Figure 10.3 Buffer overflow – manipulation of ret value.

The fundamental problem that leads to a buffer overflow attack is that there is no range/size check on the inputs received from the user, and the inputs are not parsed properly. In certain cases, buffer overflows can also lead to denial of service (DOS) attacks, as the memory might crash from the kind of input received. In other cases, the attacker gets access to unauthorized data, which leads to problems with confidentiality, or might allow them to gain elevated privilege, which might be misused. There are related problems with buffer overflow such as typecasting and integer wraparound. An integer overflow or wraparound occurs when an integer value is incremented to a value that is too large to store by the variable or if an improper value is assigned. When this occurs, the value may wrap to become a very small or negative number. This also happens when **unsigned int** to **signed int** conversions are performed in the process. Typically, controller software is written in fast languages such as C or Java. These languages do not have type-safe or memory-safe capabilities, which lead to such vulnerabilities, and it falls on the developer to be careful with the development process.

A related problem that uses crafted inputs is the **TOCTTOU** problem, which stands for Time-of-Check-to-Time-of-Use. In this case, the attacker might gain read/write access to the memory properly at a particular time, but then, the access permission is not revoked at the proper time and they end up with persistent access to resources. An example would be the "Always Sign In" options on applications, which might allow access to malicious actors instead of authorized personnel. The essential problem in the case of Unix systems is that permission checking cannot be correctly performed in user-level processes using the available Unix system calls, and the developer needs to specifically watch out for access control to prevent unauthorized access.

10.1.2 Hardware and Side-Channel Vulnerabilities

Another class of vulnerabilities is the hardware vulnerabilities. In this case, the attacker is required to have physical access to the hardware that will be compromised. There are various ways in which the attacker could use the physical access to compromise the system. A common example is the attacker using the hardware access to extract sensitive information from the system, such as cryptographic keys or other credentials. They could also use the access to understand the device better to either compromise the device in the future or use the knowledge to exploit vulnerabilities in other deployments of the device. For example, if the attacker can take apart and understand the vulnerabilities of a smart meter, they can use that knowledge to exploit other smart meters from the same vendor. The hardware analysis techniques can be grouped under two main categories – (i) reverse engineering and (ii) side-channel analysis.

Hardware reverse engineering: In this approach, the attacker could read the memory or the firmware of the device such as EEPROM, Flash/ROM, and RAM of the device to understand its operation. They could also monitor the internal communication buses such as the Serial Peripheral Interfaces (SPI) bus or Joint Test Access Group (JTAG) to understand the process workflow. Logic analyzers are used to connect to these buses to interpret the bus signal protocols by connecting them directly to the buses or test pins. To combat these issues, vendors often employ tamper-resistant manufacturing techniques, such as covering critical components with epoxy to prevent unauthorized access.

Side channel: Side channel attacks are an alternate way of obtaining unauthorized information from the hardware by using the physical access to monitor the power consumption or computation time of the system to understand its operation. These methods do not rely on brute-force techniques, as vendors take protective measures against direct compromise to the integrity of the system, such as using epoxy. The analysis that monitors the power consumption of the chip is called differential power analysis and is used to obtain information on the algorithms in the chip. The timing analysis of the chip works similarly and can be used to estimate the encryption method and keys in a chip. For example, the RSA encryption technique computes at a rate of m^e mod n, and the RSA key can be inferred based on whether a multiplication step occurs after every iteration.

10.1.3 Social Engineering

Another class of vulnerabilities are those that are exploited through a human in the loop. Humans are often the weakest link in a security system, and the attack vector leveraging on humans is called social engineering. Social engineering is attempting to gain and manipulate trust from others in order to gain access to a system or information. Several examples of social engineering attack vectors exist, such as the USB in a parking lot method. If a person were to discover a USB drive in a parking lot or on their person, they will generally plug it in. These USBs are typically infected with malware that will allow attackers gain access to critical systems. There are ongoing efforts to mitigate such attacks including training personnel to recognize suspicious behaviors and circumstances and disabling access to USB drivers and the Internet on mission critical systems. Another popular social engineering method is impersonating as a trusted third party – people will receive phone calls, text messages, or emails claiming to be the IT support desk or HR personnel.

A very popular social engineering attack is "phishing." Phishing refers to a malicious email message that attempts to pose as coming from a trusted source. There is also a variant of the general phishing attack known as spear phishing, where the attacker chooses a target, studies them carefully, and creates a custom message tailored to that specific target which leverages on personal information. For example, an employee who is expecting a message from HR confirming their promotion will be more likely to click on a suspicious email claiming to be about a revision of salary or benefits. Phishing emails are usually not authenticated or signed with the proper certificates. They also usually contain attachments with malware or try to redirect the user to malicious websites. Organizations use methods that scan attachments before delivering them to the inbox (and flag suspicious messages) and use defense mechanisms such as Proofpoint to guard against malicious URLs.

10.1.4 Malware

Vulnerabilities are usually exploited by means of malware, which comes from the portmanteau of malicious software. Malware runs of a system without the user's consent and performs malicious operations. There are different types of malware, including viruses, worms, Trojan Horses, spyware, dishonest adware, scareware, crimeware, and so on. There are three key components to malware – (i) infection/propagation method, (ii) payload, and (iii) obfuscation techniques.

Infection/propagation: This component defines how a malware attacks and spreads through a system. The initial access of a malware to a system might be through software vulnerabilities, social engineering, etc. **Shellcode** is a piece of code that exploits a vulnerability and installs the malware on a target machine. Other ways in which the malware infects a system could be through an unauthorized download onto a machine, portable media such as USB, documents with macros, or email attachments. The way to prevent infection and mitigate propagation is through hard-coded, well-known authentication mechanisms that restrict shell codes and other attack vectors that aim to install malware on a machine.

Payload: Payload refers to the actionable piece of code that executes the malicious action desired by the attacker. The payload could achieve different things such as the following:

1. *Low-level extortion*: ransomware that holds the data or application hostage in exchange for money or favors
2. *Privacy compromise*: steal financial or other confidential information
3. *Propagation*: infect other systems connected to the same network, e.g. sending spam messages from a compromised account
4. *Log user keystrokes*: obtain passwords, financial information, and other confidential information
5. *Persistent threats*: remain undetected in the system to gain insight into behaviors and become a launchpad for future attacks

A special case of a payload is a **botnet**. A botnet is a collection of computers (in the range of hundreds to thousands) running remote control software under the illegitimate control of an individual or a group. Typically, the computers perform their usual tasks for their legitimate owners before being "activated," or taken control of by the attacker. In some cases, the bots might perform both functions in parallel to avoid detection. The attacker takes remote control of the bot by using methods such as Internet Relay Chat (IRC) or other peer-to-peer software. Bots are usually used to execute attacks such as distributed denial of service (DDoS), sending spam, click fraud, or distribute a new exploit code to a large number of targets.

Disguise/obfuscation techniques: This component is responsible for disguising the malware from detection by the target. This is achieved by a variety of ways such as encryption, where the virus code is encrypted with a constant encryption technique and a variable key. The virus code could be encrypted directly on the target system or on a network command and control (C&C) communication server. The technique of using variable encryption technique and variable key is called a *polymorphic virus*.

Obfuscation refers to techniques in which the virus code is reformulated into a different piece of code, while still being functionally equivalent. This process usually involves rewriting the binaries of the attack code to avoid detection. This technique is applied *inside* the virus; hence, the obfuscated copies are not literal copies of the virus rather they behave as functional equivalents. There exist a variety of tools that the attacker can use to achieve obfuscation of their virus code. A special implementation of obfuscation techniques is *rootkits*.

Attackers would ideally like to obtain and maintain administrative (root) access on their targets without detection, even after an attack. Rootkits are pieces of code that achieve this objective. Rootkits are deployed either in the kernel mode or in the bootkit mode. In the *kernel* mode, the rootkit will modify the OS kernel or device drive to maintain access to the highest privilege level on the target. Direct Kernel Object Manipulation (DKOM) is a rootkit that directly modifies entries in the OS process table, and the scheduler is used to hide its presence. *Bootkits* is a technique that modifies the Master Boot Record (MBR) of the target system OS to obtain the highest privilege when the system boots up. Rootkits are difficult to detect and remove as they mimic the same privilege as the OS software.

10.1.5   Supply Chain

Supply chain issues can manifest in both hardware and software vulnerabilities. The modern information technology supply chain is very complex and is reliant on components coming together from different sources. These components might not necessarily come from trusted entities, and attackers can insert malicious elements in the base components that might go undetected in the final product. Software supply chain vulnerabilities include backdoors or hidden vulnerabilities. Backdoors are pieces of code that allow the manufacturer or developer to monitor the software after it has been deployed, without the knowledge of the user. An example would be the uses of third-party libraries that provide functionalities such as graphics, encryption, or logging. The malicious code can reside in different components such as the OS, services, or drivers connecting to hardware.

Hardware supply chain issues take the form of hidden hardware in the final product, which provides the attacker with capabilities of additional processing that does something other than the main purpose of the product. The hidden hardware will typically operate completely outside the control of the installed software or firmware of the main product and will not be detected in a superficial analysis. An example of hardware vulnerabilities will be a keyboard that contains an additional chip that logs keystrokes and sends them back to the manufacturer or a network switch that looks for password containing packets and periodically sends them to the attacker.

There are numerous examples of supply chain vulnerabilities impacting critical infrastructure such as the Dragonfly/Energetic Bear hack, which among other actions used batch scripts to enumerate users on a victim domain controller or added newly created accounts to the administrators group to maintain elevated access. Numerous companies have had issues with their devices being compromised in supply chains such as Cisco and HP, where the devices were shipped with malware pre-installed and these devices were then deployed in a critical infrastructure such as government bodies.

10.2 Security Mechanisms: Access Control and Malware Detection

Given the vulnerabilities discussed in the previous sections, system designers and developers have also created ways of protecting the system from these vulnerabilities. Of these, two important techniques are (i) access control and (ii) malware detection.

10.2.1 Access Control

Access control is a security mechanism that places restriction on who or what is allowed to access a resource. Access can mean to read, write, or touch the data in any way. Access control is a mechanism through which permission needs to be obtained before accessing a resource. Access control can be enforced by both hardware (processor) and software (OS). In the case of hardware-based access control, the privilege levels can be thought of as various "rings" and privilege levels are enforced at various levels. An example is shown in Figure 10.4.

The outer ring consists of user mode applications such as web browsers or document processors. Ring 2 consists of device drivers that might not directly affect operations, such as USB drivers or audio drivers. Ring 1 consists of device drivers key to operation, such as those for storage. Ring 0 is the kernel layer, which might

Figure 10.4 Hardware access control layers.

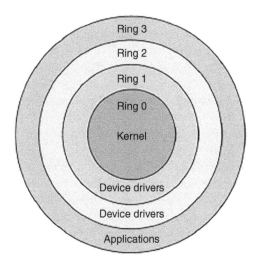

house the Windows Kernel or Linux kernel. They perform privileged functions, including memory management and BIOS. A CPU checks the instruction against the privilege level before executing the instruction. For example, a user application cannot make changes to the BIOS under normal circumstances. However, system calls and hardware interrupts sometimes allow data flow across the rings.

Access control can also be implemented through software such as the OS. In this case, the access control is based on subjects and objects. The subjects include the user and the process requesting access. Objects include files, hardware device access, etc. An access control matrix is defined, which specifies whether the subject has to read, write, or execute permissions on an object. The access control matrix can be defined using different approaches:

1. *Discretionary*: object owner determines who has access
2. *Mandatory*: security administrator determines who has access to objects
3. *Role-based*: subject's role determines their access to objects

An example of discretionary access control is the Linux file permissions system. It uses the following format to determine the access for objects: R – read, W – write, X – execute access, and the file permissions are determined by three groups – Owner, Group, or Everyone.

10.2.2 Malware Detection

The most commonly recognized mechanism of malware detection is antivirus software. Malware detection mechanisms compare programs to known malicious and malware patterns and analyze program operations for malicious operations that could lead to unsafe states. The fundamental component of all malware detection schemes is their scanning mechanism. The scanning could rely on pattern matching by comparing binaries of the software or program to known virus signatures. It could also rely on checksums – which detect if a file's integrity has been compromised to detect any changes. Another common method is referred to as "sandboxing," where the program is run in an emulated environment to verify if it either produces data that match with a virus signature or performs any actions that correspond to known malicious behavior.

Malware detection is a very challenging task because of several challenges in scanning and evaluating malicious behavior. Malware typically performs nefarious activities under mutation/obfuscation/encryption and do not exhibit their malicious patterns openly. Evaluation of malware signatures is also challenging because the companies that develop the antivirus software need to first obtain the malware from the wild before characterizing and developing its signature. This presents several challenges:

1. Only common malware patterns are usually detected by antivirus software, as these are the ones that are most analyzed and well understood.
2. Detection of new, sophisticated malware is very difficult and tends to go undetected. Newer malware that exploit zero-day vulnerabilities are not detected by most antivirus, and they need to get constant updates to ensure that the malware database is not out of date.
3. Malware developers can test their malware against the antivirus software's database, as these software need to be open to everyone.

Malware detection in industrial control systems (ICS) is different from traditional antivirus software. A traditional antivirus uses a *blacklist* for detecting and blocking malware. A control system on the other hand usually relies on a *whitelist*, which specifies all the programs that are allowed to run, while blocking others. The concept of whitelisting is also sometimes implemented in traditional IT systems such AppLocker, which is a MS Windows application that specifies rules based on a "Group policy" that controls which applications can be installed and executed. There is also the ICS Whitelist, which maintains a library of hashes of common ICS applications. This library is used to compare running programs against known valid program hashes. However, it is important to realize that whitelisting is a software construct that might have vulnerabilities of its own and is not foolproof.

10.3 Assurance and Evaluation

In detecting and mitigating vulnerabilities, an important consideration is the need to ensure that system security mechanisms are operating as designed. To assess the performance of the security mechanisms, two terms are defined – (i) assurance and (ii) evaluation.

Assurance is an estimation of the likelihood that a system will fail in a particular way. **Evaluation** is the process of assembling evidence that a system meets, or fails to meet, a prescribed assurance target.

The security mechanisms to be discussed are assessed in terms of their assurance and evaluation metrics. Security testing is done with different goals and techniques, as described in Table 10.1. It is important to understand that most security tools are "dual purpose" – they are also used by attackers to find and exploit weaknesses in target systems.

10.3.1 Port Scanning

Port scanning, sometimes referred to as system scanning, is an attempt by the attacker to discover all the open systems and ports on a network and evaluate the

Table 10.1 Goals and techniques of security testing.

Goals	Techniques
Network Discovery	Port scanning
	Network monitoring
	Network policy analysis
System vulnerabilities	Vulnerability scanning
	Continuous monitoring
Software security	Static analysis
	Dynamic analysis
	Fuzz testing

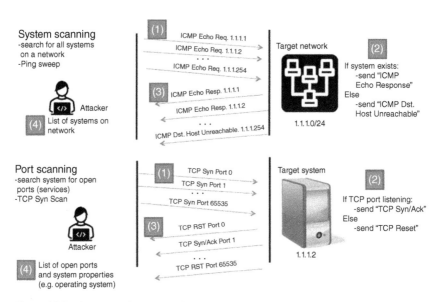

Figure 10.5 Port scanning mechanism.

system's properties. This process is usually performed using a ICMP ping message as described in Figure 10.5.

10.3.2 Network Monitoring

Network monitoring is an attempt by the attacker to search for useful information such as software banners (to discover products or versions), authentication data, or other sensitive information such as email or http traffic to help compromise the

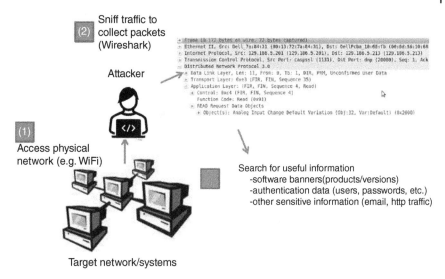

Figure 10.6 Network monitoring mechanism.

network in the future. Network monitoring is usually performed by monitoring network traffic using tools such as Wireshark. This process is shown in Figure 10.6.

10.3.3 Network Policy Analysis

Network policy is enforced by a combination of various firewalls and routers. Each component has its own ruleset and access control list that identify potential policy violations and misconfigurations. Analysis of network policy can be done either through offline review of firewall rulesets or through online probing of network to ensure compliance. Tools that are used to perform network policy analysis include NetAPT and Firemon (Figure 10.7).

10.3.4 Vulnerability Scanning

Vulnerability scanning is a method to discover potential vulnerabilities that might exist on a target system by sending a crafted message such as a malicious shellcode. The shellcode is a method to identify more serious vulnerabilities in the system such as unpatched or older versions of software, such as SSL v2. An overview of this process is shown in Figure 10.8.

10.3.5 Continuous Monitoring

Continuous monitoring is driven by the need to frequently monitor the security of systems and networks and is a major recent initiative of the US Federal

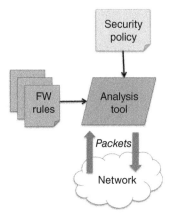

Figure 10.7 Network policy review.

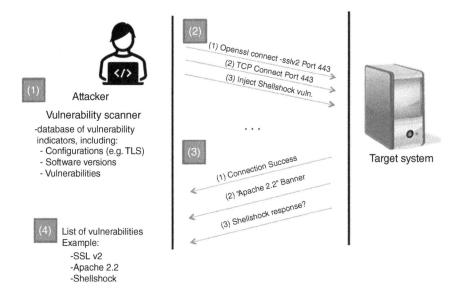

Figure 10.8 Vulnerability scanning mechanism.

Government, reflected in policy initiatives such as the NIST 800 series of guidelines. Many tools that monitor the security of systems provide incomplete information or maybe intrusive when used on live networks. "**SCAP** is the Security Content Automation Protocol that is a suite of specifications that standardize the format and nomenclature by which security software products communicate security content, particularly software flaw and security configuration information," as defined by NIST. This protocol is used to provide non-intrusive ways

Figure 10.9 Continuous monitoring using the SCAP protocol.

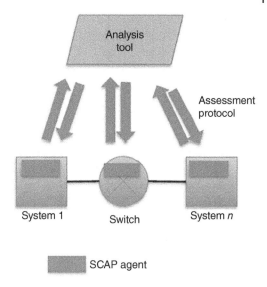

and methods to standardize continuous monitoring for critical systems. OVAL (Open Vulnerability and Assessment Language) is an XML-based language used to communicate this information. The information is then documented in databases such as CVE (Common Vulnerabilities and Exposures) and are evaluated using frameworks such as the Common Vulnerability Scoring System (CVSS) (Figure 10.9).

10.3.6 Security Assessment Concerns

An important consideration in security assessments is that many of the assessment techniques may cause issues when deployed on a live system. The tools consume bandwidth and create connections to running service that might disrupt operations. For example, a ping sweep of SCADA systems in an integrated circuit manufacturing plant caused a robotic arm to swing 180° degrees and destroy $50,000 worth of wafers. In another incident, security testing at a natural gas utility caused lockup of SCADA systems, hence shutting down the pipeline for 4 hours. Hence, extreme care is needed when deploying security assessment mechanisms to avoid disruptions to operations.

10.3.7 Software Testing

Software testing is primarily performed through either static analysis or dynamic analysis. Fuzz testing is a separate technique that is also used.

Static analysis analyzes the source code for vulnerabilities without actually executing the code. This approach has a lot of "coverage," which analyzes the code in a more *complete* manner. However, this approach might not be *sound*, as it does not execute the actual code. Static analysis evaluates many possible execution paths and thus might provide high false-positive rates. Tools that perform static analysis include Fortify, Coverify, or Splint.

On the other hand, **dynamic analysis** analyzes the code by actually executing the code and monitoring its performance. Hence, it can identify errors that might only occur at runtime. This approach is considered more *sound* than *complete*, as it actually executes each control path. It also tends to have lower false-positive rates as compared to static analysis. Tools that perform dynamic analysis include Valgrind and Purify.

Fuzz testing is a method that uses repeated randomized fault injection in a program to find a fault. Crash of the program usually signifies some sort of memory error, such as buffer overflow. Fuzz testing is performed in a controlled setting and not on a live/operational system. The inputs that or randomized faults can be in the form of (i) faulty network packets, (ii) improper file formats, or (iii) shared memory compromises. Fuzz testing is not foolproof and is usually poor at code coverage or completeness. Tools such as JBroFuzz or SPIKE are used for fuzz testing. Several fuzz testers exist for smart grid protocols such as IEC 61850 and DNP3.

10.3.8 Evaluation

A combination of security mechanisms is required to protect ICS devices from malicious activity. A protection "profile" is created for a system that defines the security requirements around the various components such as the (i) OS, (ii) access control systems, (iii) boundary control devices, (iv) intrusion detection systems, (v) smartcards/hardware validation methods, (vi) key management systems, and (vii) VPN client. The profile might also define requirements for other components as necessary.

A device is tested for the safety of this safety profile, and the performance is evaluated. There are several Evaluation Assurance Levels (EALs), which define the extent to which testing was performed. EALs have several levels:

1. *EAL1*: Functionally tested, which has no detailed analysis.
2. *EAL 4*: Methodically designed, tested, and reviewed. An example of this is various OS such as Red Hat Linux, Windows 7, etc.
3. *EAL 6*: Semiformally verified design and tested. An example of a software passing EAL6 is Green Hills OS.
4. *EAL7*: Formally verified design and tested. These are usually custom-made applications.

10.4 Compliance: Industrial Practice to Implement NERC CIP

The only standard mandated for compliance for the smart grid is the NERC Critical Infrastructure Protection (CIP) standard. This standard is specifically for the Bulk Energy System (BES), although additions are being proposed for distribution systems with grid edge devices, such as IoT devices. The NERC CIP standard was formulated by NERC and is ratified by FERC, with the current version being Version 5, which has been effective since 1 April 2016. NERC can issue fines of up to $1 million/day for non-compliance of the CIP standard. The evaluation for compliance is based on periodic assessments of a utility's security and is performed by an auditor.

Version 5 of the standard introduced the "Bright line criteria" – which is used to classify cyber assets based on violation risk factor (VRF). VRF is defined over three levels:

1. *High*: Most control centers for reliability coordinators, balancing authorities, and generator operating centers fall under this category
2. *Medium*: Transmission facilities over 500 kV or 200–499 lV with interconnection to greater than two lines are included in this category
3. *Low*: All other control system devices are included in this category

NER CIP lays out several requirements for these different classes of assets and has a specific guideline for protecting assets against cyber threats in CIP-007.

10.5 Summary

This chapter describes vulnerabilities that can be present in the components used in the smart grid including software and hardware vulnerabilities, side-channel vulnerabilities, and social engineering vulnerabilities. Various security mechanisms that are deployed against these vulnerabilities are discussed, primarily access control and malware detection. The assurance and evaluation of these security mechanisms and their methodology have been discussed. Finally, compliance requirements for devices used in the smart grid are introduced.

10.6 Problems

1 Which of the following is a side-channel analysis technique against a cryptosystem?

A Brute force analysis
B Rubber hose cryptography
C Differential power analysis
D Cryptanalysis

2 What are the common examples of social engineering?
   A Phishing
   B Software exploits
   C IP spoofing
   D Malware

3 A buffer-overflow overwrites what stack value to hijack program flow?
   A Return address
   B Frame pointer
   C Function variables
   D Unused memory

4 What is "whitelisting" in malware detection for OT systems?
   A Verifying the system/program behavior against known malicious patterns
   B Exclusion of known malware by comparing from a database
   C Allow only specific users to run programs
   D Allow only certain "applications" to run on a system

5 What is "assurance" in software testing?
   A Guarantees provided by the developer on system performance
   B Guarantees provided by the operating system on application reliability
   C Estimation of the likelihood that a system will continue to perform in face of problems
   D Estimation of the likelihood that a system will fail in a particular manner

10.7 Questions

(1) What are software vulnerabilities? What are some of the most relevant software-related failures in the smart grid?
(2) What are the key components of a malware? Explain in detail.
(3) How is access control used for mitigation of attacks? Explain the levels of access control.
(4) What are the different types of software testing? Explain in detail.
(5) What are NERC CIP guidelines? Explain how NERC CIP guidelines will be used to protect a generator at a generation station.

Further Reading

Aloul, F., Al-Ali, A.R., Al-Dalky, R. et al. (2012). Smart grid security: threats, vulnerabilities and solutions. *International Journal of Smart Grid and Clean Energy* 1 (1): 1–6.

Clark, A. and Zonouz, S. (2017). Cyber-physical resilience: definition and assessment metric. *IEEE Transactions on Smart Grid* 10 (2): 1671–1684.

Francia, G.A. III and El-Sheikh, E. (2022). NERC CIP standards: review, compliance, and training. *Global Perspectives on Information Security Regulations: Compliance, Controls, and Assurance*, pp. 48–71.

Khurana, H., Hadley, M., Lu, N., and Frincke, D.A. (2010). Smart-grid security issues. *IEEE Security and Privacy* 8 (1): 81–85.

Venkataramanan, V., Srivastava, A.K., Hahn, A., and Zonouz, S. (2019). Measuring and enhancing microgrid resiliency against cyber threats. *IEEE Transactions on Industry Applications* 55 (6): 6303–6312.

Zhang, Y., Yau, D., Zonouz, S. et al. (2017). Guest editorial smart grid cyber-physical security. *IEEE Transactions on Smart Grid* 8 (5): 2409–2410. https://doi.org/10.1109/TSG.2017.2735244.

11

Smart Grid Case Studies

This chapter deals with various smart grid case studies – deployments, challenges, successes, and learning. We will get familiar with the future smart grid model, learn about various smart grid use cases, and study the risks that come with increased "smartness" in the grid.

The next-generation electrical power system will be typified by the increased use of communications, information technology, and computation technology in the generation, delivery, and consumption of electrical energy. This is shown in the conceptual model in Figure 11.1.

11.1 Smart Grid Demonstration Projects

Various smart grid projects have been demonstrated and executed over the past 20 years. Smart grid projects are driven by various factors such as follows:

1. Economic and business case
2. Social adoption and benefits
3. Legal and regulatory compliance
4. Environmental consideration – climate change, and other related concerns
5. Human resources and challenges in staffing leading to more automation
6. Advanced technical solutions

Various smart grid projects have been deployed over the years. It is challenging to identify the first smart grid project as the definition of "smartness" used by various deployments vary quite widely. In general, smart grid projects define "smartness" as characterized by the use of advanced communication and computation to deliver power in the most efficient way possible. This may also include efforts to gain more situational awareness, which lead to smarter decisions. In general, it is accepted that the concept of smart grids started gaining traction around the turn of the century, with the deployment of a Wide-Area Monitoring System (WAMS) by

Cyber Infrastructure for the Smart Electric Grid, First Edition.
Anurag K. Srivastava, Venkatesh Venkataramanan, and Carl Hauser.
© 2023 John Wiley & Sons Ltd. Published 2023 by John Wiley & Sons Ltd.

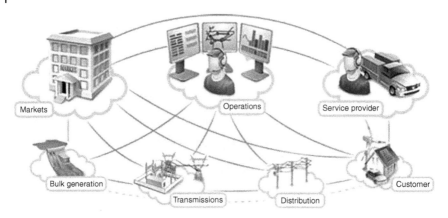

Figure 11.1 Conceptual model of the future smart grid.

the Bonneville Power Administration (BPA) considered as a watershed event. The earliest smart grid projects usually involved either deployment of smart meters or other advanced metering infrastructure (AMI) or the deployment of WAMS infrastructure. A short summary of a few early, pioneering projects is described below.

Enel: Enel S.p.A. is an Italian utility and is one of the world's largest utilities and power companies. Enel deployed a smart grid project called the Telegestore project, and it was highly unusual in the utility world because the company designed and manufactured their own smart meters, aggregated the data by itself, and developed their own system software for the meters and situational awareness. This project is considered to be one of the largest and among the first smart grid projects providing more data on residential networks to the system operators.

GridWise: The GridWise Alliance is an initiative that began at the Pacific Northwest National Lab (PNNL) and is an alliance between technologists, stakeholders, and the government that aims to bring fundamental developments to grid operations through advanced information technology, motivated by the philosophy that "bits are cheaper than iron." GridWise is involved in several smart grid projects such as the Pacific Northwest Smart Grid Demonstration Project.

Utilities and smart grid: Avista Utilities in Spokane, Washington, deployed a smart grid project that involved installation of new communications, and monitoring infrastructure aimed toward increasing the efficiency of operation by installing automated switches and inclusion of additional measurements in the Distribution Management Software (DMS) to reduce the need for truck visits for system maintenance and diagnosis. Florida Power and Light, a utility serving the state of Florida, commissioned a smart grid project that installed over 3 million smart meters, 250 home area networks (HAN), 9000 intelligent distribution devices, 45 phasor measurement units (PMUs), and advanced monitoring equipment in over 270 substations. In 2009, the American Recovery and Reinvestment Act of 2009 awarded federal stimulus to several smart grid projects including ISO New England, which installed multiple PMUs and phase data concentrators (PDCs) to reduce wide-scale blackouts and increase the reliability.

Smart cities: In addition to utilities and regulators, several cities have commissioned smart city projects, the earliest of which were in Boulder, Colorado, Austin, Texas, and Chattanooga, Tennessee. These cities installed several million smart meters as part of a wider effort to provide better situational awareness to the operators and increase reliability to consumers. Several examples of smart cities are also present in Europe, such as Eindhoven in the Netherlands. Currently, smart cities are being developed across the globe in all continents.

In general, smart grid projects can be broadly considered to be in the following areas:

1. Advanced metering infrastructure and customer systems including (i) peak demand reductions and electricity savings; (ii) meter operations and maintenance savings; (iii) consumer behavior studies; (iv) impact of advanced metering, dynamic pricing, enabling technologies, and information treatments on consumer behavior; and (v) effectiveness of marketing, consumer education, and outreach programs.
2. Distribution systems projects including (i) impact on reliability, (ii) energy efficiency improvements (e.g. lower line losses), and (iii) operations and maintenance savings.
3. Transmission system projects including (i) applications of synchrophasor technologies and systems, (ii) energy storage systems, and (iii) technical and financial performance.
4. Marketplace innovation including (i) introduction of new products and services and (ii) electric vehicle and distributed energy resources integration incentives
5. Cybersecurity projects, such as advancements in cybersecurity practices.

11.2 Smart Grid Metrics

Smart grid projects need careful planning and deployment, as upgrades to critical infrastructure are expensive and can negatively affect a large number of people if things go wrong. Smart grid projects hence use different metrics to quantify progress and help in planning out expenditure. These projects typically develop a Metrics and Benefits Reporting Plan, which are documents that describe the smart grid assets, functions, impacts, and related data that will be collected by the recipients and reported to United States Department of Energy. Each project has its own unique metrics and benefits that are pre-determined. As the projects are implemented, equipment installed, and functions made operational, utilities or the project recipients track two types of smart grid metrics – (i) build metrics and (ii) impact metrics. Build metrics measure progress toward deployment of smart grid assets. Build metrics include, for example, reports from recipients on the number of smart meters installed, the number of substation automated, and the number of dynamic pricing programs offered. Impact metrics measure how, and to what extent, these smart grid assets are affecting grid operations and performance or how they enable customer programs and behavior changes. Impact metrics include, for example, reports from recipients on the magnitude of peak demand reductions, the number of truck rolls reduced, and the amount of maintenance cost avoided as a result of the projects smart grid activities.

Other examples of smart grid metrics include the following:

1. Percentage of customers capable of receiving information from grid operators and the percentage of customers opting to make or delegate decisions about electricity consumption based on that information.
2. Percentage of distributed generation and storage devices that can be controlled in coordination with the needs of the power system.
3. The number of smart grid products for sale that have been certified for end-to-end interoperability.
4. The number of measurement points per customer for collecting data on power quality, including events and disturbances.
5. The amount of distributed generation capacity (MW) that is connected to the electric distribution system and is available to system operators as a dispatchable resource.
6. The percentage of grid assets (e.g. transmission and distribution equipment) that are monitored, controlled, or automated.
7. The percentage of entities that exhibit progressively mature characteristics of resilient behavior.

11.3   Smart Grid Challenges: Attack Case Studies

While the smart grid increases the efficiency of operations, it also presents various challenges – primarily with the increasing attack surface for malicious actors. In this section, we will study two major incidents that illustrate the challenges with the increasing attack surface.

11.3.1 Stuxnet

Stuxnet attack was a computer worm discovered in June 2010 that spreads via Microsoft Windows. Its primary target was toward Industrial Control Systems (ICSs), specifically aimed at Siemens SCADA systems. The overall attack included several steps:

1. Zero-day exploits,
2. A Windows rootkit,
3. The first ever programmable logic controller (PLC) rootkit,
4. Antivirus evasion techniques,
5. Complex process injection and hooking code,
6. Network infection routines,
7. Peer-to-peer updates, and
8. A command and control (C&C) interface.

Stuxnet had several variants but broadly used a Windows common vulnerabilities and exposures (CVE) to gain initial access. This was done using CVE-2010-2568, a Windows Shell vulnerability in Microsoft Windows XP SP3, Server 2003 SP2, Vista SP1 and SP2, Server 2008 SP2 and R2, and Windows 7, allowing local users or remote attackers to execute arbitrary code via a crafted (i) .LNK or (ii) .PIF shortcut file, which is not properly handled during icon display in Windows Explorer. The vulnerability could allow remote code execution if the icon of a specially crafted shortcut is displayed. The steps that the LNK vulnerability used to gain access are shown in Figure 11.2. An attacker who successfully exploited this vulnerability could gain the same user rights as the local user. Once initial access is gained, the attacker would then use CVE-2010-2772, in Siemens Simatic WinCC and PCS 7 SCADA system. It used a hard-coded password, which allows local users to access a backend database and gain privileges on the controller's HMI. Once the worm has infected a host machine, it used network propagation techniques through the MS10-061 Print Spooler zero-day vulnerability to communicated with peers on the network and propagate to additional machines. It used remote procedure calls (RPCs) to communicate with additional hosts on the network to ensure that all the machines have access to the latest version of the Stuxnet worm.

Once the Stuxnet worm has been successfully installed, it periodically ran checks to ensure that it received updates from its C&C server. This step is illustrated in Figure 11.3. The Stuxnet worm hooks specific APIs that are used to open project files inside the **s7tgtopx.exe** process, which when loaded executed the Stuxnet files. The worm would then manipulate the statement list (controller files) within the PLC to install the malware on the PLC memory. The malware monitors the Profibus messaging bus of the system and modifies the frequency

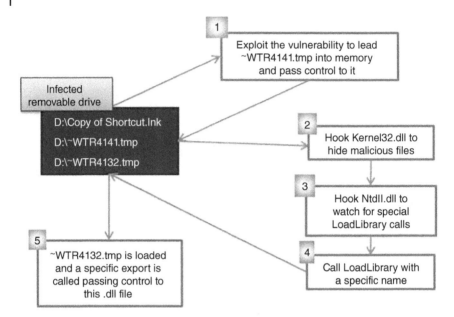

Figure 11.2 LNK vulnerability – from infected removable drive to gaining elevated privilege on the host.

of the variable speed drives from 1410 to 2 Hz and then to 1064 Hz, which affects the operation of the connected motors by changing their rotational speed. In addition, it installs a rootkit that masks the changes in rotational speed from the monitoring systems. A major target of the Stuxnet attack was the Natanz nuclear facility in Iran, which in November 2010 ceased operations due to several technical problems associated with the centrifuge's operational capacity.

11.3.2 Ukraine Attack

The Ukraine cyber attack exploited a Windows vulnerability defined in CVE-2014-4114. The CVE allows "remote attackers to execute arbitrary code via a crafted OLE object in an Office document." This is not a zero-day vulnerability, meaning that the operator can already know that this vulnerability exists in their system. The steps of the attack are also shown in Figure 11.4. A simplified attack timeline for the Ukraine attack is given as follows:

1. Phishing emails are sent to substation operators containing the BlackEnergy malware
2. The operator opens the Office document that enables the attacker to exploit CVE-2014-4114 and install malware on the IT network computer

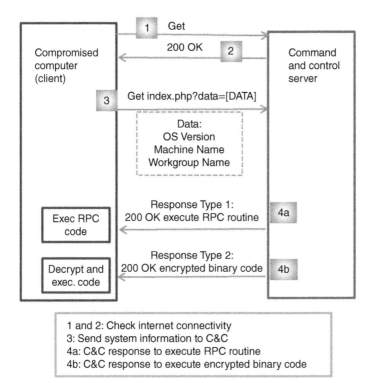

Figure 11.3 Infected host communicating with command and control server to receive updates.

3. The attacker gains access to credentials for the VPN into the OT network
4. The attacker gains unauthorized elevated privileges on the substation HMI computer
5. The attack is coordinated between multiple substations to maximize the impact
6. The attacker opens the switches in the system and disable remote access by the operator

Following these steps, an attacker can isolate the critical loads in the system from the sources and will also prevent the operator from taking mitigative control actions unless the attack has been detected immediately after the vulnerability is exploited. If the exploit is not detected immediately, the substation operator does not have access to the monitoring tools that provide awareness. The operator is not aware of the attack as the attacker accesses the OT network using valid credentials. The operator becomes aware after the attacker gains unauthorized elevated privileges, and there is a physical system impact due to the circuit breaker

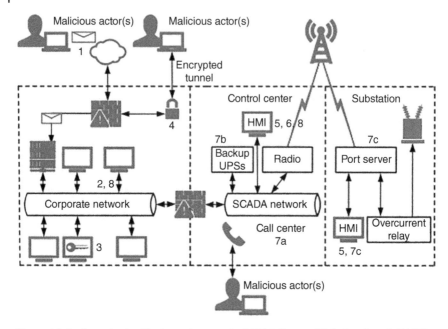

Figure 11.4 Steps in the Ukraine cyber attack of 2018. Source: Whitehead et al. [2017].

opening. At this point, the operator is forced to take drastic control actions such as performing hard reset on the system to restore system operation.

11.4 Mitigation Using NIST Cybersecurity Framework

In the United States, NIST provides a structure and grouping for security controls to be deployed to protect the smart grid, which it calls as the "Security Control Identifiers and Family Names." There exist a wide variety of security tools available for the defenders to protect their systems, and there are various frameworks proposed that dictate how these technologies need to be deployed. Not all parts of the control system require the same level or type of security. The NIST cybersecurity framework has a comprehensive set of security controls which provides guidance to determine and develop a defense-in-depth approach for ICS systems. This defense-in-depth method has been proposed by the United States Department of Homeland Security and many other researchers and represents the best practice adopted by various utilities across the world. The defense-in-depth approach as implemented by Schweitzer Engineering Laboratories (SEL) is shown in Figure 11.5.

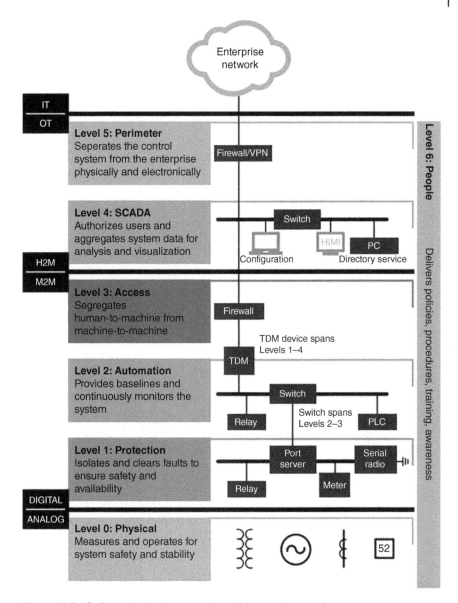

Figure 11.5 Defense-in-depth approach to mitigate cybersecurity concerns.
Source: Whitehead et al. [2017].

Different methods of security controls deployed at appropriate levels allow the user to efficiently monitor, detect, and deter attempts to circumvent the security. The system is demarcated at various points such as human-to-machine (H2M) interaction with the system (laptop or workstation), application of products in the system (HMI, RTU, relay), and communications protocols (SCADA or protection protocols). Under this security model, users design cybersecurity defenses and power system operation controls and settings in such a way that devices can continue to operate properly even if they are isolated from the network in the case of cyber attacks.

A layered security approach is considered to be the best practice for the smart grid, with a customized level of security defense mechanisms deployed for each type of control system device. Good cyber security has to be a combination of people, hardware, software, policies, and procedures, regardless of it is deployed for an enterprise network or an ICS. The Ukraine cyber incident and the Stuxnet attack were events that demonstrated how cyber attacks could affect critical infrastructure, which were previously considered to be immune to such attacks. There is an increased focus on the nature of cyber-physical systems and the need to protect critical infrastructure in the future.

11.5 Summary

In this chapter, we studied various smart grid deployments, their economic and business case. We looked at some pioneering smart grid efforts, and typical types of smart grid upgrade projects. The metrics and procedures for planning smart grid projects were discussed, following a detailed look at two pivotal cybersecurity events – the Stuxnet attack and the Ukraine attack. A detailed description of these events was provided, with a discussion on mitigation mechanisms that are deployed in the smart grid that addresses these threats.

11.6 Problems

1 Which factor does NOT motivate new smart grid projects?
   A Economic or business case
   B Legal or compliance requirement
   C Environmental factors
   D Staffing challenges leading to automation
   E Keeping with other critical infrastructure domains such as water or transportation systems

2 Which statement is false?

A Stuxnet exploited multiple zero-day vulnerabilities

B Stuxnet attacked systems from certain vendors only

C A network physically separated from the Internet can be infiltrated by Stuxnet

D The author of Stuxnet created two fake certificates to sign drivers

3 Which of the following is NOT considered a smart grid project?

A Deployment of AMI

B Integration of advanced controllers in the system

C Deployment of cybersecurity tools for protection of system

D Training of staff on new technologies such as cloud computing

E Market solution innovation by integrating new services

4 Which of the following malware was used in the Ukraine cyber attack of 2018?

A BlackEnergy

B Triton

C Stuxnet

D Industroyer

E Pipedream

11.7 Questions

(1) What are some of the metrics used to measure the impact of smart grid projects?

(2) The article "W32.Duqu: The precursor to the next Stuxnet" describes Duqu, a threat similar to Stuxnet. Compare Stuxnet and Duqu from the following aspects: (a) initial infection, (b) propagation, and (c) command and control.

(3) Describe the Ukraine cyber attack from 2018 in detail. What measures can be taken to avoid a repeat of this incident?

(4) From the reference " Defense-in-Depth Security for Industrial Control Systems" from Schweitzer Engineering Labs (SEL) discuss the defense-in-depth framework. How does it help protecting the smart grid?

Further Reading

Belhomme, R., Tranchita, C., Vu, A. et al. (2011). Overview and goals of the clusters of smart grid demonstration projects in France. *2011 IEEE Power and Energy Society General Meeting*, pp. 1–8. IEEE.

Bossart, S.J. and Bean, J.E. (2011). Metrics and benefits analysis and challenges for Smart Grid field projects. *IEEE 2011 EnergyTech*, pp. 1–5. https://doi.org/10.1109/EnergyTech.2011.5948539.

Hashmi, M., Hänninen, S., and Mäki, K. (2011). Survey of smart grid concepts, architectures, and technological demonstrations worldwide. *2011 IEEE PES Conference on Innovative Smart Grid Technologies Latin America (ISGT LA)*, pp. 1–7. https://doi.org/10.1109/ISGT-LA.2011.6083192.

Jiang, Y., Liu, C.-C., Diedesch, M. et al. (2016). Outage management of distribution systems incorporating information from smart meters. *IEEE Transactions on Power Systems* 31 (5): 4144–4154. https://doi.org/10.1109/TPWRS.2015.2503341.

Mallet, P., Granstrom, P.-O., Hallberg, P. et al. (2014). Power to the people!: European perspectives on the future of electric distribution. *IEEE Power and Energy Magazine* 12 (2): 51–64. https://doi.org/10.1109/MPE.2013.2294512.

Marnay, C., Aki, H., Hirose, K. et al. (2015). Japan's Pivot to resilience: how two microgrids fared after the 2011 earthquake. *IEEE Power and Energy Magazine* 13 (3): 44–57. https://doi.org/10.1109/MPE.2015.2397333.

Melton, R. (2015). Pacific Northwest Smart Grid Demonstration Project Technology Performance Report Volume 1: Technology Performance. *No. PNW-SGDP-TPR-Vol. 1-Rev. 1.0; PNWD-4438*, Volume 1. Richland, WA (United States): Pacific Northwest National Lab. (PNNL).

Mulder, W., Kumpavat, K., Faasen, C. et al. (2012). Global inventory and analysis of smart grid demonstration projects.

Skjølsvold, T.M. and Ryghaug, M. (2015). Embedding smart energy technology in built environments: a comparative study of four smart grid demonstration projects. *Indoor and Built Environment* 24 (7): 878–890.

Whitehead, D.E., Owens, K., Gammel, D., and Smith, J. (2017). Ukraine cyber-induced power outage: analysis and practical mitigation strategies. *2017 70th Annual Conference for Protective Relay Engineers (CPRE)*, pp. 1–8. IEEE.

Index

a

access control   159–160
  overview   131
  RBAC in IEC 62351   131–132
accountability   117
address resolution protocol (ARP)   77
  spoofing attack   139
Advanced Encryption Standard (AES)
    120
advanced metering infrastructure (AMI)
    24
advanced persistent threats (APT)   114
aggregation   101
aggressive   130
anomaly-based detection   147
anomaly-based IDS   146
application layer   75–76
Application Service Data Units (ASDUs)
    129
associations   101
assurance and evaluation   161
  continuous monitoring   163–165
  evaluation   166
  network monitoring   162–163
  network policy analysis   163
  port scanning   161–162
  security assessment concerns   165
  software testing   165–166
  vulnerability scanning   163

asymmetric encryption   119
  cryptographic protocols   122–123
  elliptic curve cryptography   121
  hash functions   122
asynchronous transfer mode (ATM)   64
auction market   97
authentication   117–118
  asymmetric   119
    cryptographic protocols   122–123
    elliptic curve cryptography   121
    hash functions   122
  certificates   125
  digital signatures   124
  Kerckhoffs's *versus* Kirchoff's Law
      118–120
  message authentication codes
      123–124
  symmetric key encryption   119,
      120–121
Authentication Header (AH)   143
authenticity   117
automatic generation control (AGC)
    28
availability   116

b

balancing authority (BA)   36–37, 97
best-effort IP   72
bilateral contracts market   97

Cyber Infrastructure for the Smart Electric Grid, First Edition.
Anurag K. Srivastava, Venkatesh Venkataramanan, and Carl Hauser.
© 2023 John Wiley & Sons Ltd. Published 2023 by John Wiley & Sons Ltd.

Bonneville Power Administration (BPA) 172

bootkits 158

Border Gateway Protocol (BGP) 140

botnet 157

broadcast 68–70

brute force attack 119

business management system (BMS) 27

c

C37.118 87–89

cascading faults 106

certificate authorities (CAs) 144

certificate generation, encryption 125

challenge–response mode 130

circuit breakers and switches 9

class 101

client server model 59

cloud computing 107–109

Common Information Model CIM (IEC 6170) 103

Common Smart Inverter Profile (CSIP) 91

Common Vulnerability Scoring System (CVSS) 165

communication links 58

communication systems

  data loss and corruption 52–53

  jitter 50

  multi-hop networks 51–52

  processing delay 51

  propagation delay 47

  queuing delay 49–51

  transmission delay 47–49

composition 101

confidentiality 116

confidentiality, integrity, and availability (CIA) 116–117

Connectionless-mode Network Service (CLNS) 78

consumer data 117, 118

continuation power flow 7

continuous monitoring 163–165

control centers

  CIM (IEC 6170) 103

  conventional control centers 95–97

  future control centers 98–99

  IEC 61850 103–105

  modern control centers 97–98

  RDF 103

  smart grid

    continuation power flow 29

    functions of 29

    state estimation 28

    unit commitment and economic dispatch 29–30

  UML 100–102

  XML 102–103

conventional control centers 95–97

core security properties

  confidentiality, integrity, and availability 116–117

  encryption and authentication 117–123

  privacy and consumer data 117

COVID-19 pandemic 4

crafted packet 137

Critical Infrastructure Protection (CIP) 38

cryptanalysis 120

cryptographic protocols 122–123

cryptography

  DNP3 secure authentication 129–130

  IEC 62351 128

  power system communications 127

  principles and threats 118–120

cyber-physical system 1, 12

cybersecurity

  defense-in-depth approach 178–179

  risk 115–116

threats   114
vulnerabilities   114–115
cybersecurity, smart grid   30–31

d

data modeling, IEC 61850   104
data storage   104
decentralized architecture   98
denial-of-service (DoS)   135–136
   DDoS   138
   flooding   136–137
   malformed packet   137
   reflection   137–138
Diffie–Hellman   121
digital signatures   124
Direct Kernel Object Manipulation
   (DKOM)   158
Discrete Fourier transform (DFT)   22
discretionary access control (DAC)   131
disguise techniques   157
distributed denial of service (DDoS)
   138
distributed energy resources (DERs)   2,
   91
Distributed Network Protocol 3 (DNP3)
   protocols   83–85
distribution management systems (DMS)
   96, 172
DNP3 secure authentication   129–130
domain name system (DNS)   75, 77–78,
   136, 140
"downgrade" attacks   144
dynamic analysis   166
Dynamic Host Configuration Protocol
   (DHCP)   67

e

Eastern interconnection   35
"Echo Request" messages   136
electric power grid
   components of   1–2

control system   5–6
   defined   1
   electricity flow structure   4
   interconnections   2–3
   power grid operation   6–8
   voltage levels in   5
Electric Reliability Council of Texas
   (ERCOT)   35
electric reliability organization (ERO)
   37
elliptic curve cryptography   121
encapsulated security payload (ESP)
   122, 143, 144
encryption   117–118
   asymmetric   119
   cryptographic protocols
      122–123
   elliptic curve cryptography
      121
   hash functions   122
   certificates   125
   digital signatures   124
   Kerckhoffs's *versus* Kirchoff's Law
      118–120
   message authentication codes
      123–124
   symmetric key encryption   119,
      120–121
end systems   58
Enel   172
Energy Independence and Security Act
   of 2007 (EISA)   40
energy management system (EMS)   19,
   27, 81, 95
error correcting codes   52
error detection codes   52
ethernet   62–64, 128
Evaluation Assurance Levels (EALs)
   166
Extensible Markup Language (XML)
   102–103

f

fast decoupled load flow (FDLF) method
    7
fault-tolerant computing   105–107
field devices   83
firewalls, network protection   144–145
FirstEnergy (FE)   11–12
flooding   135, 136–137
fuses   9
future control centers   98–99
fuzz testing   166

g

Gauss–Seidel technique   7
Generic Object Oriented Substation
    Events (GOOSE)   70, 86, 128
Generic Substation Events (GSE)   104
Generic Substation State Events (GSSE)
    86
Global Positioning System (GPS)-based
    clocks   92
global positioning system (GPS) signal
    23
glue protocols
    DNS   77–78
    link layer address   76
GridWise Alliance   172
Group Domain of Interpretation (GDOI)
    128, 144

h

hardware access control layers   159
hardware reverse engineering   155
hardware vulnerability   155
hash functions, encryption   122
high-voltage direct current (HVDC)   2
home area networks (HAN)   173
host-based intrusion detection system
    (HIDS)   146
hosts   58
HTTP flood   136

Human–Machine Interface (HMI)
    82–83
hybrid   146

i

ICMP flood   136
IEC 61850   86–87, 103–105
IEC 62351   128
IEEE 1588   23. *see also* Precision Time
    Protocol (PTP)
IEEE C37.118 data format   89
independent system operators (ISO)   36
Independent System Operators (ISOs)
    97
industrial control systems (ICS)   161,
    175
infection/propagation method   156–157
Information and Communication
    Technology (ICT) systems   95
infrastructure as a service (IaaS)   109
inheritance   101
initial random sequence numbers (ISNs)
    140
insufficient entropy   120
integrity   116
intelligent electronic devices (IEDs)   48,
    70, 83, 97
interconnection, North American power
    grid   2–3
Inter-Control Center Communications
    Protocol (ICCP)   87, 88
Internet Corporation for Assigned
    Names and Numbers (ICANN)
    68
internet protocol (IP)   65, 138
    addressing   66–68
    datagram format   65–66
Internet Security Association and Key
    Management Protocol (ISAKMP)
    123, 143
internet service providers (ISPs)   68

intrusion detection system (IDS)
145–146
anomaly-based detection   147
signature-based detection   147–148
IPSec   122, 143–144
IPSec Key Exchange (IKE)   123

j

jitter   50
Joint Test Access Group (JTAG)   155

k

Kerckhoffs's *versus* Kirchoff's Law
118–120
kernel mode   158
Kirchoff's Law   118–120

l

LAN address   76–77
layered communication model
application layer   75–76
glue protocols
DNS   77–78
link layer address   76
link layer, service models
ethernet   62–64
network layer
broadcast and multicast
68–70
IP addressing   66–68
routing   68
OSI   59–60
physical layer   60–61
TCP/IP models   59–60, 78
transport layer
TCP   73–74
UDP   72
lines of code (LoC)   152
link layer address   76
link layer services   61
link virtualization   63–64

LNK vulnerability   176
Load Serving Entities (LSEs)   97

m

malformed packet   135, 137
malware   156–158
malware detection   160–161
man-in-the-middle (MITM) attack   140
Manufacturing Messaging Specification
(MMS)   86
Master Boot Record (MBR)   158
master terminal unit (MTU)   83
message authentication codes (MACs)
61, 76–77, 123–124
meter data management system (MDMS)
25
mode-of-operation   120
modern control centers   97–98
multicast   68–70
multi-hop networks   51–52
multiprotocol label switching (MPLS)
63

n

National Energy Reliability Corporation
(NERC)   97
National Institute of Standards and
Technology (NIST)   40
NERC Critical Infrastructure Protection
(CIP) standard   167
network   114
network attacks and protection
denial-of-service   135–136
DDoS   138
flooding   136–137
malformed packet   137
reflection   137–138
intrusion detection   145–146
anomaly-based detection   147
signature-based detection   147–148

network attacks and protection (*contd.*)
  network protection through firewalls 144–145
  network protection through security protocols
    IPsec 143–144
    TLS 141–143
  spoofing
    ARP Spoofing 139
    other spoofing 139–140
network-based intrusion detection system (NIDS) 146
Network Control Program (NCP) 57
network interface card (NIC) 62
network layer
  broadcast and multicast 68–70
  IP addressing 66–68
  routing 68
network monitoring mechanism 162–163
network policy 163, 164
network protection
  through firewalls 144–145
  through security protocols
    IPsec 143–144
    TLS 141–143
Network Time Protocol (NTP) 92
Newton–Raphson method 7
NIST cybersecurity framework 178–180
nodes 58, 61
North American Electric Reliability Corporation (NERC) 30, 37–38
North American power grid, interconnection 2–3

o
obfuscation techniques 157
Object Management Group (OMG) 100
object-oriented (OO) programming 100–101

open systems interconnection (OSI) layer 59–60
Open Vulnerability and Assessment Language (OVAL) 165
operational structure, smart grid
  rest of the world 40–41
  standards and interoperability 39–40
  system reliability 37–39
operator 82

p
Pacific Northwest National Lab (PNNL) 172
packet duplication strategies 69
payload method 157
peer-to-peer model 59
phase data concentrators (PDCs) 173
phase measurement unit (PMU)
  applications 23–24
  architecture 20
  data packets 23
  faster synchronized data 19
  phasor calculation 22
  phasor estimation 21–22
  sampling rate of 21
  synchronization time signal 22–23
phasor data concentrator (PDC) 58, 88
phasor measurement unit 58, 88
physical layer 60–61
"Ping of Death" attack 137
platform-as-a-service (Paas) 109
point-to-point protocol. *see* transmission control protocol (TCP)
polymorphic virus 157
port 71
port scanning mechanism 161–162
power grid operation
  applications 6–8
  blackouts 10–12
  infrastructure 8–9
Precision Time Protocol (PTP) 92

privacy   117
processing delay   51
propagation delay   46, 47
protocol-based attacks   135
protocols   59

q
queuing delay   46, 49–51

r
rate of change of frequency (ROCOF)
    20
reflection attack   137–138
Regional Transmission Operators (RTOs)
    97
regional transmission organizations
    (RTOs)   36
remote terminal unit (RTU)   83
reporting schemes, IEC 61850   104
Resource Description Framework (RDF)
    103
RFC 5905. *see* Network Time Protocol
    (NTP)
risk, cybersecurity   115–116
Rivest, Shamir, Adleman (RSA)   121
role-based access control (RBAC)   131
    in IEC 62351   131–132
rootkits   157, 158
routers   58, 65
routing   68, 69

s
Sampled Measured Values (SMV/SV)
    86
sandboxing   160
secure authentication (SA)   129–130
Secure Socket Layer (SSL)   141
security assessments   165
security associations (SAs)   123, 143
security-constrained economic dispatch
    (SCED)   27, 30, 96

security-constrained unit commitment
    (SCUC)   27, 29
security mechanisms
    access control   159–160
    malware detection   160–161
security test   162
sequence of events (SOE)   27
Serial Peripheral Interfaces (SPI)   155
Shellcode   156
side-channel vulnerability   155
signature-based detection   147–148
signature-based IDS   146
smart cities   173
Smart Energy Profile (SEP)   91
smarter electric grid   12–13
smart grid
    communication infrastructure in   26
    computational infrastructure
        components   26–27
        control center applications   28–30
    cybersecurity in   30–31
    demonstration projects   171–173
    electric power grid
        overview   1–6
        power grid operation   6–8
    future conceptual model   172
    metrics   173–174
    NIST cybersecurity framework
        178–180
    sense   18
        phase measurement unit   19–24
        smart meters   24–26
    smarter electric grid   12–13
    standards and interoperability   39–40
    Stuxnet attack   175–176
    Ukraine cyber attack   176–178
    utilities   172–173
Smart Grid Architecture Model (SGAM)
    framework   40
Smart Grid Interoperability Panel (SGIP)
    40

smart metering   89–91
smart meters
    advanced metering infrastructure
        24–25
    communication systems for   25–26
"smartness" grid   171
Snort IDS   148
social engineering   155–156
socket   71
software   115
software-as-a-service (SaaS)   109
software-defined networks (SDN)   75
software testing   165–166
software vulnerability   120, 152–154
specification-based IDS   146
spoofing   138–140
Standard Generalized Markup Language
        (SGML)   102
static analysis   166
Stuxnet attack   175–176
Substation Configuration Language
        (SCL)   104
supervisory control and data acquisition
        (SCADA)   5, 26–27, 129, 165
    application layer protocols   84
    C37.118   87–89
    distributed energy resources   91
    ICCP   87
    protocols
        components   82–83
        DNP3   83–85
        IEC 61850   86–87
    smart metering   89–91
    system architecture   82
    time synchronization   92
supply chain   158–159
switches   58
symmetric key encryption   119, 120–121
systems   115
system scanning   161–162
system vulnerability
    hardware and side-channel   155

malware   156–158
social engineering   155–156
software vulnerabilities   152–154
supply chain   158–159

t
taxonomy of faults   108
TCP/IP   128
    models   59–60, 78
    security enhancements   141
TCP Syn flood   136
TCP/UDP segment format   72
"Teardrop" attack   137
threats, cybersecurity   114
Time-of-Check-to-Time-of-Use
        (TOCTTOU)   154
time synchronization (TimeSync)   86,
        92
transmission control protocol (TCP)
        73–74
transmission delay   47–49
transmission rate   45, 48
transmission time   45
transport layer
    TCP   73–74
    UDP   72
Transport Layer Security (TLS)   128,
        141–143

u
Ukraine cyber attack   176–178
Unified Markup Language (UML)
        100–102
user datagram protocol (UDP)   72, 136
U.S. Federal Energy Regulatory
        Commission (FERC)   37–38

v
violation risk factor (VRF)   167
virtual private networks (VPNs)   143
"Volt-Var control" mode   91

vulnerability 114–115
vulnerability scanning mechanism 163,
    164

w
weighted least squares (WLS) technique
    8
Western interconnection 35
wide-area monitoring and control
    (WAMPAC) 58

Wide-Area Monitoring System (WAMS)
    171–172

x
XML 102–103
XML Schema Definition (XSD)
    103

z
"zero-day" attacks 151